Liposomes - Advances and Perspectives

Edited by Angel Catala

Published in London, United Kingdom

IntechOpen

Supporting open minds since 2005

Liposomes - Advances and Perspectives
http://dx.doi.org/10.5772/intechopen.77926
Edited by Angel Catala

Contributors
Angel Catala, Rui Krause, Alain Bapolisi, Christian Nkanga, Okafor Nnamdi, Anibal Disalvo, Maria Frias, Roberto López-Rendón, Ketzasmin A. Terrón-Mejía, Inocencio Higuera-Ciapara, Javier Hernandez, Evelin Martínez-Benavidez, Alina Porfire, Marcela Achim, Cristina Barbalata, Iulia Rus, Ioan Tomuta, Cecilia Cristea

First published in London, United Kingdom, 2019 by IntechOpen
IntechOpen is the global imprint of INTECHOPEN LIMITED, registered in England and Wales, registration number: 11086078, The Shard, 25th floor, 32 London Bridge Street
London, SE19SG - United Kingdom
Printed in Croatia

British Library Cataloguing-in-Publication Data
A catalogue record for this book is available from the British Library

Additional hard and PDF copies can be obtained from orders@intechopen.com

Liposomes - Advances and Perspectives
Edited by Angel Catala
p. cm.
Print ISBN 978-1-78984-494-8
Online ISBN 978-1-78984-495-5
eBook (PDF) ISBN 978-1-83881-232-4

For EU product safety concerns:
IN TECH d.o.o., Prolaz Marije Krucifikse Kozulić 3, 51000 Rijeka, Croatia, info@intechopen.com or visit our website at intechopen.com.

Meet the editor

Angel Catalá was born in Rodeo (San Juan, Argentina). He studied chemistry at the Universidad Nacional de La Plata, Argentina, where he received a PhD degree in Chemistry (Biological Branch) in 1965. From 1964 to 1974, he worked as an assistant in biochemistry at the School of Medicine, Universidad Nacional de La Plata, Argentina. From 1974 to 1976, he was a Fellow of the National Institutes of Health (NIH) at the University of Connecticut, Health Center, USA. From 1985 to 2004, he served as a full professor of biochemistry at the Universidad Nacional de La Plata, Argentina. He is a member of the National Research Council (CONICET), Argentina, and Argentine Society for Biochemistry and Molecular Biology. For many years, his laboratory has been interested in the lipid peroxidation of biological membranes from various tissues and different species. Professor Catalá has directed 12 doctoral theses and published over 100 papers in peer-reviewed journals, several chapters in books, and 12 edited books. He has received awards at the 40th International Conference on Biochemistry of Lipids 1999, Dijon (France). He was the winner of the Bimbo Pan-American Nutrition, Food Science and Technology Award 2006 and 2012, South America, Human Nutrition, Professional Category. In 2006, he received the Award in Pharmacology, Bernardo Houssay, in recognition of his meritorious works of research. Angel Catalá belongs to the editorial board of the *Journal of Lipids*, *International Review of Biophysical Chemistry*, *Frontiers in Membrane Physiology and Biophysics*, *World Journal of Experimental Medicine and Biochemistry Research International*, *World Journal of Biological Chemistry*, *Oxidative Medicine and Cellular Longevity*, *Diabetes and the Pancreas*, *International Journal of Chronic Diseases & Therapy*, and *International Journal of Nutrition*. He is also the coeditor of the *Open Biology* journal.

Contents

Preface

Liposomes have received increased attention in recent years. Nevertheless, liposomes, due to their various forms and applications, require further investigation. These structures can deliver both hydrophilic and hydrophobic drugs. Preparation of liposomes results in different properties for these systems. In addition, there are many factors and difficulties that affect the development of liposome drug delivery structures.

The purpose of this book is to concentrate on recent developments in liposomes. The articles collected in this book are contributions by invited researchers with long-standing experience in different research areas. We hope that the material presented here is understandable to a broad audience, not only scientists but also people with a general background in many different biological sciences. This volume offers up-to-date, expert reviews of the fast-moving field of liposomes and is divided in two major sections encompassing four chapters.

In the first chapter, Dr. Català describes liposomes advances and perspectives. In Chapter 2, lipid hydration and bilayer properties are described by Dr. Disalvo. General perception of liposomes and their formation, manufacturing, and applications are summarized by Dr. Krause Rui in Chapter 3. In Chapter 4, Dr. López-Rendón describes dissipative particle dynamics simulations of self-assemblies of liposomes for drug delivery applications, and finally Dr. Porfire describes the pharmaceutical development of liposomes using the QbD approach.

I would like to express my gratitude to Mr. Luka Cvjetkovic, the publishing process manager, and Intech Open Access publisher for their efforts in the publishing process.

Angel Català
Facultad de Ciencias Exactas,
Instituto de Investigaciones Fisicoquímicas Teóricas y Aplicadas
(INIFTA-CCT La Plata-CONICET),
Universidad Nacional de La Plata,
La Plata, Argentina

Section 1

Introduction

Chapter 1

Introductory Chapter: Liposomes - Advances and Perspectives - My Point of View

Angel Catala

1. Introduction

Liposomes are vesicular arrangements composed of one or more phospholipid bilayers surrounding an aqueous core. Liposomes were discovered almost six decades ago. Due to its versatility, liposomes are now analyzed for their applicability both in laboratory techniques and in medical studies. Its interest lies in its ability to traverse cell membranes and to transport certain types of molecules to defined places in the human body.

Liposomes can carry both hydrophilic and hydrophobic molecules. The preparation of the liposomes results in different properties for these systems. There are several factors involved in the preparation of liposomes that can modify their structures. Due to its biological compatibility, nonimmunogenicity, greater solubility of chemotherapeutic agents, and its ability to encapsulate a wide variety of drugs, the supply of drugs using liposomes has meant a great advance. The purpose of this book is to focus on recent developments in liposomes. The chapters selected in this book are contributions from invited researchers with long experience in different areas of research. This book offers expert and updated reviews of the field of liposomes.

2. Brief history of liposomes

Liposomes were discovered in 1961 by Alec Bangham, a British scientist who studied blood coagulation. Bangham and RW Horne were testing the new electron microscope of the Agricultural Research Council Institute of Animal Physiology, Babraham, Cambridge, England, when they observed for the first time applying the negative staining technique for the study of the structures of the lipid phase that the dispersions of lecithin contained spherulites composed of concentric sheets [1]. These images served as the first evidence that the cell membrane was a two-layered lipid structure. The width of the lipid layer was estimated at 44 Å. Preparations of phosphatidylcholine-cholesterol of equal molar ratios were described as basically the same as phosphatidylcholine alone.

A year later, in a Cambridge pub, Weissmann in a discussion with Bangham called these structures "liposomes" in honor of the lysosome, a simple organelle whose latency linked to the structure could be interrupted with detergents and streptolysins. [2], and that his laboratory had been studying: liposomes can be easily distinguished from micelles and hexagonal lipid phases by transmission electron microscopy by negative staining [3].

 IntechOpen

3. My participation in studies with liposomes

Forty-four years ago, as an international scholar of the NIH in the Department of Biochemistry of the Health Center of the University of Connecticut, I carried out studies related to the mechanism of stearoyl-CoA desaturase [4]. That is where I first prepared liposomes by sonication of egg lecithin or dimyristoyl lecithin.

Since then I have used liposomes in multiple studies in order to analyze: the exchange of palmitic acid from cytosolic proteins to microsomes, mitochondria, and lipid vesicles [4]; the oleic acid transfer from microsomes to egg lecithin liposomes [5]; the interaction of albumin and fatty-acid-binding protein with membranes: oleic acid dissociation [6]; the removal of fatty acids but not phospholipids from microsomes liposomes and sonicated vesicles by fatty-acid-binding protein [7]; Fe^{2+}- and Fe^{3+}-initiated peroxidation of sonicated and nonsonicated liposomes made of retinal lipids in different aqueous media [8]; lipid peroxidation of membrane phospholipids in the vertebrate retina [9]; the antioxidant properties of melatonin and structural analogues on Fe(2+)-initiated peroxidation of sonicated liposomes made of retinal lipids [10]; the antioxidant behavior of melatonin and structural analogues during lipid peroxidation [11]; and the use of soybean phosphatidylcholine liposomes as model membranes to study lipid peroxidation photoinduced by pterin [12].

4. General remarks, conclusions, and perspectives

It has been fascinating to follow the field of liposomes research during almost five decades. From my experience, it is impossible to predict which aspects in liposomes research will dominate in the future.

Acknowledgements

The author is grateful to his mentors: Rodolfo Brenner and Phillip Strittmatter, all his Ph.D. students, and postdocs.

Author details

Angel Catala[†]
Instituto de Investigaciones Fisicoquímicas Teóricas y Aplicadas, (INIFTA-CCT La Plata-CONICET), Facultad de Ciencias Exactas, Universidad Nacional de La Plata, Argentina

*Address all correspondence to: catala@inifta.unlp.edu.ar

† The author is a member of Consejo Nacional de Investigaciones Científicas y Técnicas (CONICET) Argentina

References

[1] Bangham AD, Horne RW. Negative staining of phospholipids and their structural modification by surface-active agents as observed in the electron microscope. Journal of Molecular Biology. 1964;**8**:660-668

[2] Sessa G, Weissmann G. Incorporation of lysozyme into liposomes: A model for structure-linked latency. The Journal of Biological Chemistry. 1970;**245**:3295-3301

[3] Yash Roy RC. "Lamellar dispersion and phase separation of chloroplast membrane lipids by negative staining electron microscopy" (PDF). Journal of Biosciences. 1990;**15**:93-98

[4] Avanzati B, Catalá A. Exchange of palmitic acid from cytosolic proteins to microsomes, mitochondria and lipid vesicles. Acta Physiologica Latino Americana. 1982;**32**:267-276

[5] Catalá A, Avanzati B. Oleic acid transfer from microsomes to egg lecithin liposomes: Participation of fatty acid binding protein. Lipids. 1983;**18**:803-807

[6] Catalá A. The interaction of albumin and fatty-acid-binding protein with membranes: Oleic acid dissociation. Archives Internationales de Physiologie et de Biochimie. 1984;**92**:255-261

[7] Zanetti R, Catalá A. Fatty acid binding protein removes fatty acids but not phospholipids from microsomes liposomes and sonicated vesicles. Molecular and Cellular Biochemistry. 1991;**100**(1):8

[8] Fagali N, Catalá A. Fe^{2+} and Fe^{3+} initiated peroxidation of sonicated and non-sonicated liposomes made of retinal lipids in different aqueous media. Chemistry and Physics of Lipids. 2009;**159**:88-94

[9] Catala A. Lipid peroxidation of membrane phospholipids in the vertebrate retina. Frontiers in Bioscience (Scholar Edition). 2011;**3**:52-60

[10] Fagali N, Catalá A. Melatonin and structural analogues do not possess antioxidant properties on Fe(2+)-initiated peroxidation of sonicated liposomes made of retinal lipids. Chemistry and Physics of Lipids. 2011;**64**:688-695

[11] Fagali N, Catalá A. The antioxidant behaviour of melatonin and structural analogues during lipid peroxidation depends not only on their functional groups but also on the assay system. Biochemical and Biophysical Research Communications. 2012;**423**:873-877

[12] Thomas AH, Catalá Á, Vignoni M. Soybean phosphatidylcholine liposomes as model membranes to study lipid peroxidation photoinduced by pterin. Biochimica et Biophysica Acta. 2016;**1858**:139-145

Section 2

Liposomes General Properties

Chapter 2

The Role of Water in the Responsive Properties in Lipid Interphase of Biomimetic Systems

Anibal Disalvo and Maria de los Angeles Frias

Abstract

The lack of details in the hydration properties of lipid bilayers hinders the design of biomimetic systems that, as liposomes and vesicles, may be used for biotechnological and medical purposes. In this chapter, studies indicate water as a membrane dynamic component determining the affinity and response of lipid membranes to amino acids, peptides and others stimuli. Based on thermodynamic analysis in lipid monolayers and its comparison with swelling shrinkage processes in liposomes and vesicles, it is concluded that: (1) the interphase of a lipid bilayer in a bidimensional solution of hydrated polar groups imbibed in labile water can be exchanged with the media by osmosis and or expansion-compression. (2) Excess water beyond the hydration shell (confined water) has solvent properties for additives in the bulk water phase and confers free energy that is in excess for binding of amino acids and peptides. (3) Dissolution in the water membrane phase changes the water activity (a_w) and affects the surface pressure. (4) Defects may be formed by the compression of bilayers in which carbonyl groups organized water differently. These studies indicate that a deeper understanding of the role of lipid bilayers in cellular biology and support the development of future lipid-based biotechnology that should necessarily include the role of water as a membrane dynamic component.

Keywords: lipid bilayers, hydration, osmotic stress, responsive membranes, curvature, defects

1. Introduction: cell membranes and lipid membranes

Cell membranes are complex systems composed by lipids and proteins [1, 2]. They have structural and functional properties that are essential for life. The lipid bilayer is the backbone of cell membranes and is mostly composed by amphiphilic molecules such as phospholipids, cholesterol and glycolipids among others [3–7] (**Figure 1**).

A qualitative step in describing the membrane properties in terms of the lipid composition was the isolation and purification of lipids and its stabilization in water (**Figure 1**-Part B). After Bangham's discovery in the 1960s [8, 9], lipids were found to form closed particles (liposomes) that are able to trap in controlled conditions such as ions, macromolecules and polar molecules of different nature in the inner aqueous space (**Figure 2**). With this information, it was thought reliable the possibility to modulate such trapping properties by changing membrane composition and to orient filled liposomes to specific organ targets. The changes were in the

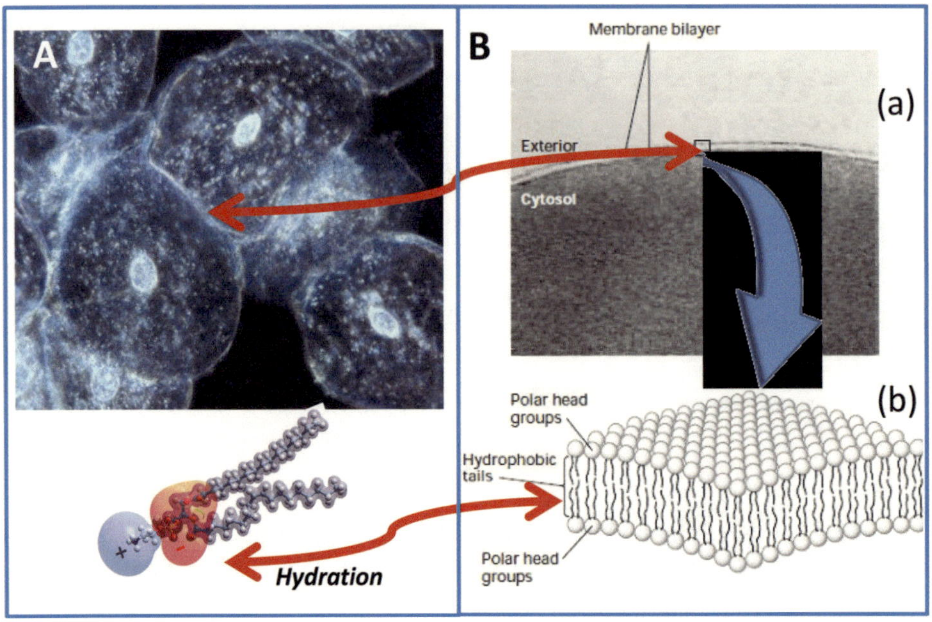

Figure 1.
(A) Electronic microscopic of cells; (B) cell membrane backbone is the lipid bilayer (a) which is formed by amphiphilic compounds such as lipids (single phospholipid) dispersed in water (b) facing the polar head group to water and segregating the acyl chains.

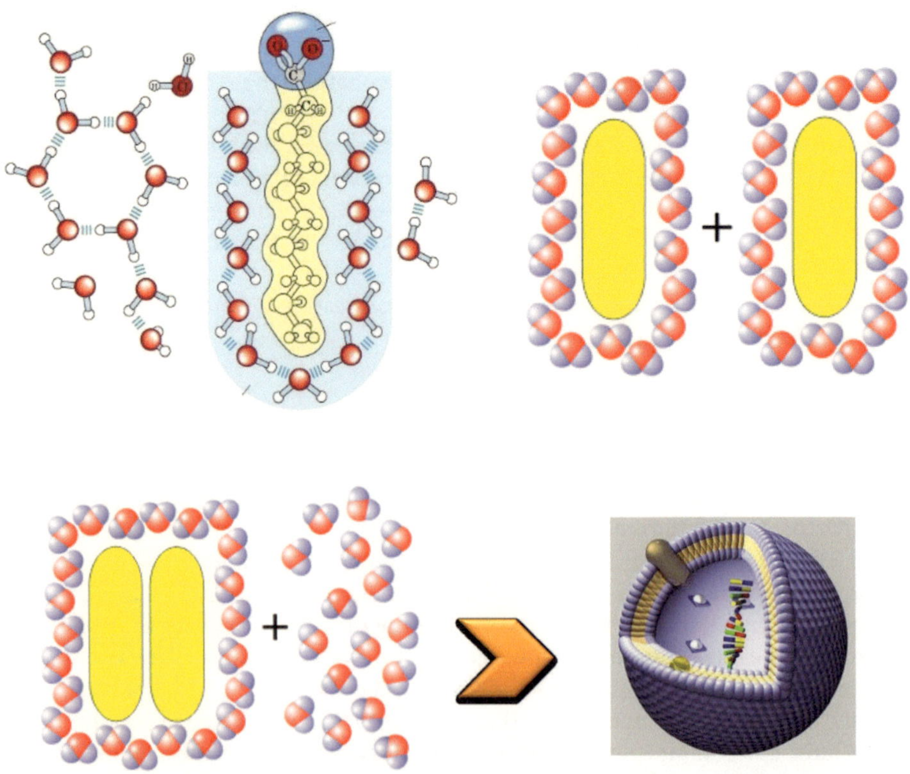

Figure 2.
Organized water around the acyl chains is displaced promoting aggregation by an increase in entropy of the system (upper part). Lipids self-aggregate in water due to the hydrophobic interaction of the nonpolar chains forming closed particles such as liposomes (lower part).

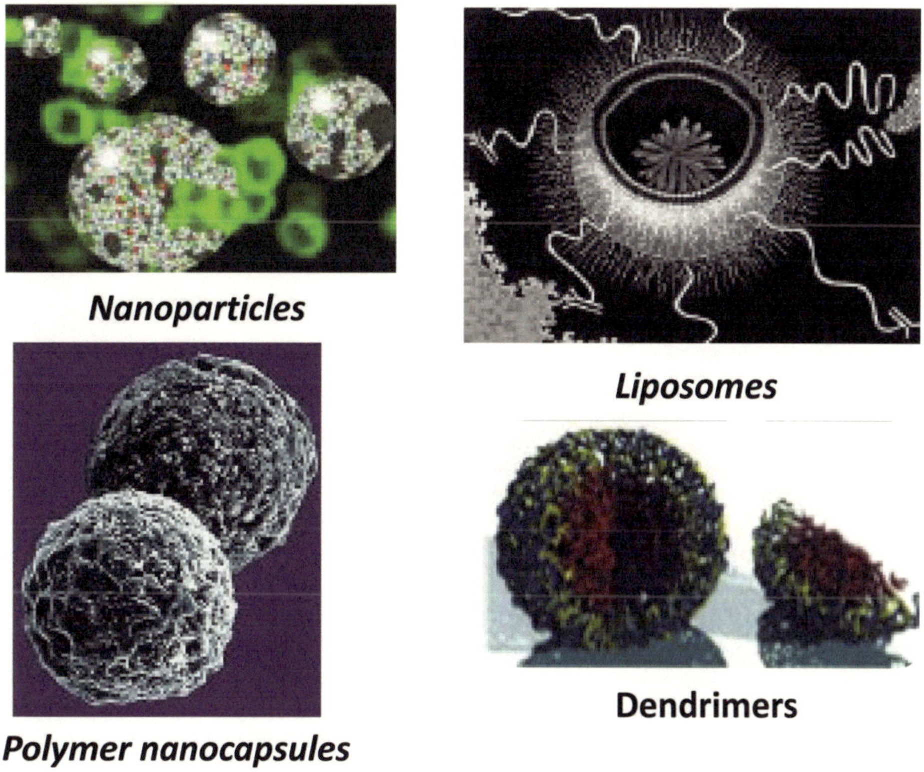

Figure 3.
Liposomes are one of the most attractive biomimetic systems because its preparation is done with lipids extracted from cells. In addition, other biomimetic nanoparticles can include lipids in its matrix.

direction to modulate the surface properties, the excluded aqueous volume, the water permeability and the mechano-elastic properties of the particles.

In addition, different methodologies have been designed, afterwards, in order to obtain suspension of homogeneous size distribution of different magnitudes [10–12]. With this wide range of possibilities, it was immediate to infer that liposomes and its different versions of covered or uncovered unilamellar vesicles would be the ideal tools to trap, vehiculate specific compounds to drive them to specific targets and deliver drugs to organs and tissues, specifically for human beings pathologies (**Figure 3**).

However, there are a number of difficulties for the direct use of these preparations that are mostly derived by limited knowledge of the physicochemical properties of the lipid bilayers. These include the presence of water as a major component in the membrane matrix; the thermodynamic properties derived from it in relation to the response from physicochemical stimuli and the interphase properties.

The purpose of this chapter is to analyze these points in order to propose new strategies for designing biomimetic lipid particles more efficiently.

2. Lipid hydration and bilayer stabilization

When dry phosphatidylcholines (PCs) of chain length above 12 hydrocarbon atoms are dispersed in water above their transition temperature as described in **Figure 2**, they form lamellar onion-like structures in which bilayers are separated by aqueous spaces that are available to trap the compounds of interest to vehiculize and deliver (**Figure 4**).

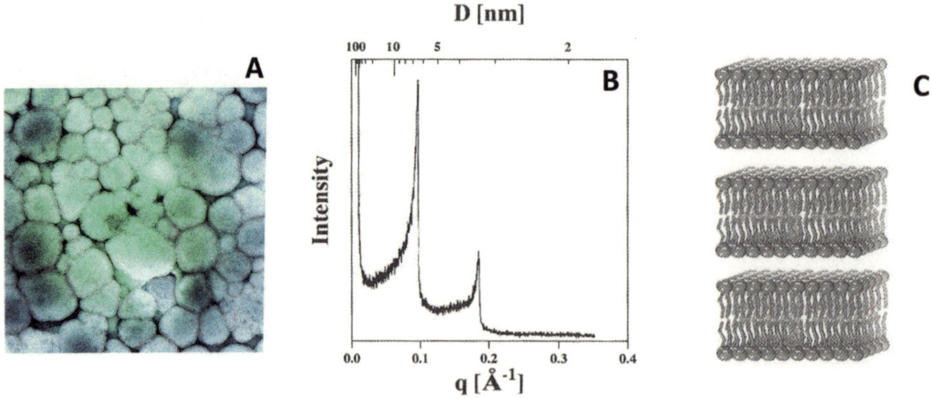

Figure 4.
(A) Electronic microscopic traditional image of multilamellar liposomes; (B) the diffraction pattern illustrates the separation between bilayers; (C) water solution trapped in between bilayers is schematically represented.

After Bangham, Luzzati and others [13–15] determined the thickness of lipid bilayers by SAXS. Different thermal profiles were obtained according to the lipid features and head group structure was found [16–18]. Lipids stabilize differently according to its geometry. Phosphatidylcholines, that have similar areas in the head group region and the acyl chain, form bilayers by stacking molecules in cylindrical shape. In contrast, phosphatidylethanolamine (PE) forms hexagonal phases due to the conical shape of molecules [19, 20].

The process of lipid hydration that derives the formation of liposomes consists of different stages as described in **Figure 5**.

The hydration step consists of the increase in area and decrease in thickness up to around 20% c.a. 22–24 water molecules per lipid. After this stage, area and thickness remain constant and the swelling of liposomes starts by the increase of water in the interlamellar space. This description of lipid swelling illustrates about several structural and physicochemical properties of the bilayers. The first observation is that there is a defined number of water molecules per lipid that determines the area per lipid and the bilayer thickness. This number is around 7–8 below the phase transition temperature and 22–24 above as derived by DSC [17, 18, 21, 22]. Thus, water is a component of the structure of the lipid bilayer, determining its thermodynamic stability.

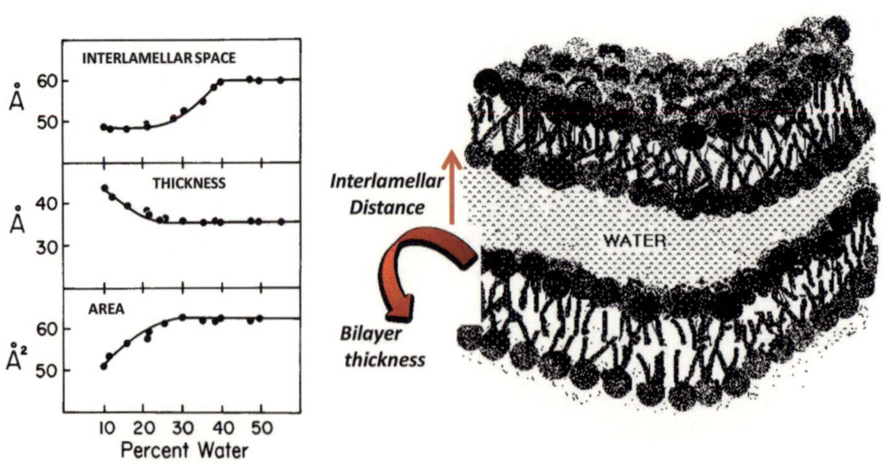

Figure 5.
Hydration of lipids and swelling of liposomes.

At least four features deserve discussion. At equilibrium in fully hydrated state, the membrane thickness is composed by the excluded volume of the lipid molecules and the excluded volume of the water organized by them (the hydration number denoted above). Thus, the barrier properties do not only merge with the head group and the acyl chain region per se but also of the packing and arrangement of water molecules in the hydration shell of the phospholipids. This means that for any solute, to overcome the bilayers, that is, releasing the trapped solute or incorporating some of them must permeate or alter the hydration shell. This can, in principle, be done by some of these mechanisms: insertion in the water network removing or replacing water molecules in the hydration shell or changing the area per lipid by expansion or compression.

The second feature of lipid bilayers is derived from the first one. The interbilayer space consists of water not bound to the membrane, that is, it can be displaced by changing the osmotic gradient between the inner spaces and the outer media of the liposome. Water can permeate the lipid bilayer with certain facility depending on the phase state of the lipids, the presence of double bonds or ramifications in the acyl chains [23, 24]. In contrast, membrane is completely or partially impermeable to some solutes, such as sugars, ions, depending on its size and molecular structure. The differences in permeation rates between water and any of these solutes means that at least in the beginning of the process, a gradient of water chemical potential can be built with a difference in solute concentrations between the two sides of the bilayer. Let us consider the interbilayer space described in **Figure 5**. If solute is more concentrated between the bilayers, water will be driven to enter due to a difference in osmotic pressure and them the spacing (and hence the trapped volume) will be larger. An opposite effect can be caused, if the solute is concentrated in the outer media of the liposomes. In this case, liposomes shrink and the interbilayer space decreases.

The third feature is represented by an area per lipid for a membrane thickness of 60 Å, when lipids hydrate with 22–24 water molecules per lipid. Thus, any change in the number of these water molecules will affect thickness and area with concomitant effect on permeability.

Finally, the fourth feature is defined by the limit of the volume decrease. This is given by the steric repulsion of the groups in the surface of the bilayer, in which water plays a significant role. Water associated with the lipids is oriented at the bilayer surface constituting an electrical potential that hinder bilayers approach. This repulsive force is named as dipole potential or hydration forces [24–27] (**Figure 6**).

It is immediate to derive that the presence of these forces hinders the adhesion or fusion of membranes with different kinds of surfaces (inorganic and organic materials, other membranes, proteins, tissues, etc.). On the other way round, those processes will be feasible if water of hydration is totally or partially removed. This point is essential to understand the role of water in terms of membrane response to biologically relevant effectors.

Entering details of the dynamic properties of membranes in relation to water, we may again inspect **Figure 5**. *The equilibrium point* corresponding to an area per molecule of 75 Å and 20 water molecules per lipid could be modified by the inclusion or exclusion of water molecules. So, the question is which perturbations can trigger *changes in hydration* that can be dominated in order to promote controlled changes in permeation. In this direction, let us focus on the mechanical and chemical forces at constant temperature. By mechanical forces, we refer to processes that led the membrane to expand or contract and by chemical forces, the competition with water by membrane sites of compounds may form hydrogen bonds. To analyze these points, previous considerations about water as a component of membranes must be done.

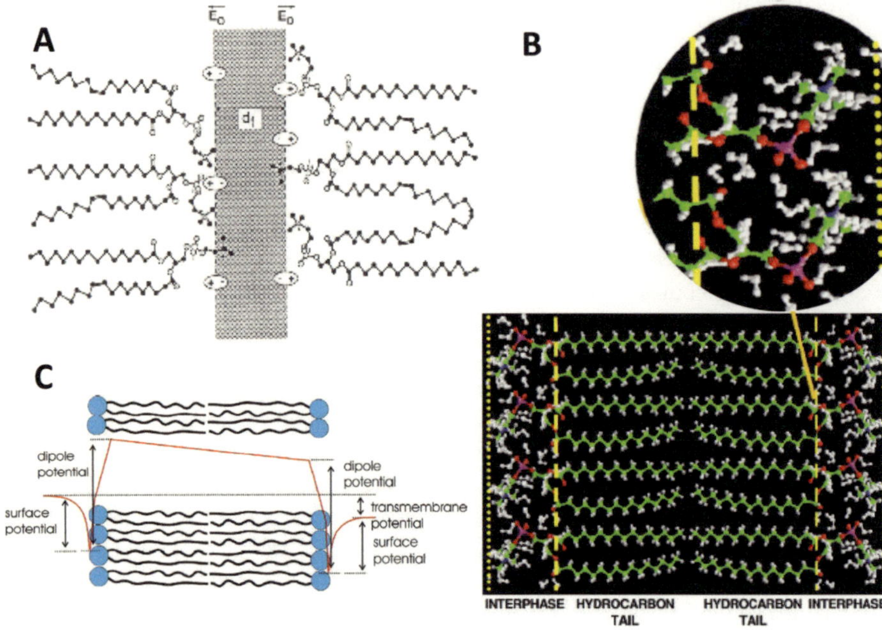

Figure 6.
(A) The limit of approach of lipid bilayers, (B) water organized at the interphase determining the repulsion forces, (C) dipole potential at the lipid interphase. This potential makes the bilayer interior positive and has important consequences in the binding and penetration of charged peptides and proteins.

3. The bilayer structure and the water ratio

The understanding of the structural role of water in lipid membranes received considerable attention after Luzzatti et al. and efforts were addressed to give a defined location to water and to determine the appropriated values for area per lipid molecule [22, 27, 28]. Today, it is generally accepted that PCs admit up to a limit of around 22–24 water molecules per lipid above the phase transition temperature to stabilize in a bilayer with an area per lipid of 64 $Å^2$ and a thickness of around 40 A [29, 30]. In the first stages of the hydration process, the water molecules interact with the phosphate group and, at a second step, with the carbonyl groups [31–33]. The formation of hydrogen bonds between these groups and water molecules can be monitored by observing the frequency shift of their stretching frequencies [34, 35]. It is expected that upon hydrogen-bonding, these frequencies decrease as a consequence of the elongation of the chemical bonds in the relative functional groups. Furthermore, the conformation of the acyl chains changes with the hydration [36].

At least three regions of differential hydration can be identified according to the rate of exchange between membrane phase and the aqueous environment: the phosphate group, the carbonyl groups and the hydrocarbon chains. Each of them has different level of water organization in regard to coordination water by H bonding [30]. From them, CO and acyl chains seem to be important to evaluate the role of water in regulatory and functional response [37, 38].

Figure 7 shows the lipid bilayer denoting the water region (upper part figure) and the profile of water distribution along the lipid bilayer (lower part).

It is observed that water covers all the phosphate and choline regions and partially the carbonyl groups. Moreover, it is also observed that water may penetrate beyond the carbonyl region, that is, the first methylene groups of the hydrocarbon chains. This denotes that different arrangements of water can be found along the membrane thickness each of one with a different energetic profile, that is, the ability

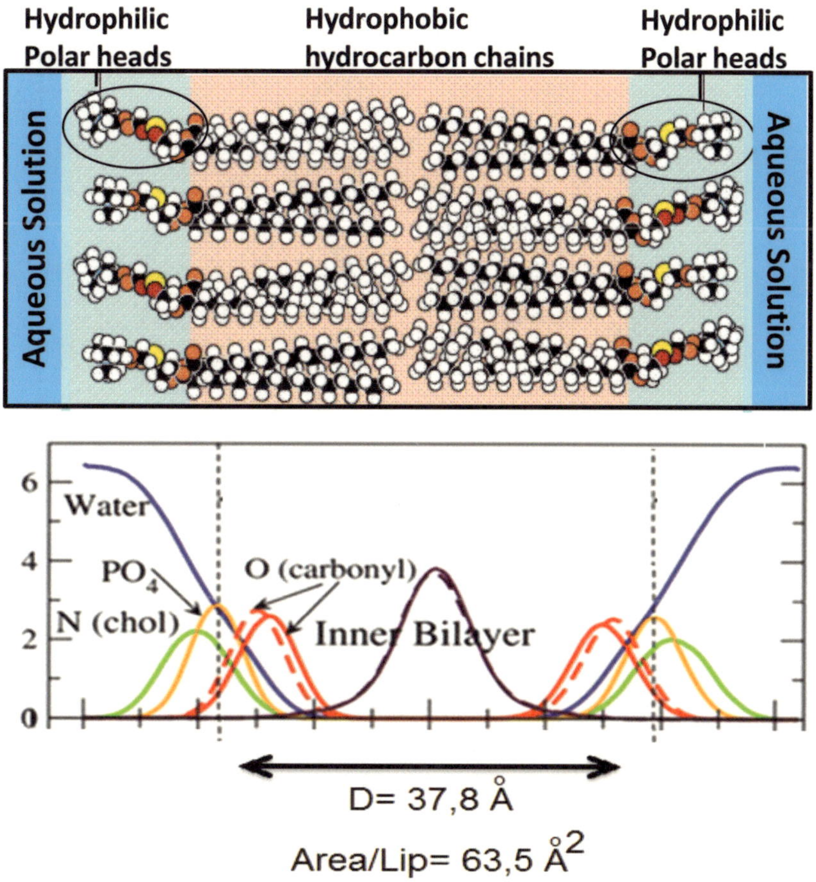

Figure 7.
The lipid bilayer considering the water interphases. In the lower diagram, it is observed that the water profile (blue line) penetrates up to the region of the carbonyl groups (red solid and dashed lines), being phosphates (orange line) and cholines (green line) groups completely covered by water.

to react with polar or ionic solutes. This reactivity can be ascribed to the residual capacity the water molecules have to form H bonds depending on the surface groups with which its interacts.

In **Figure 8**, water populations are visualized using infrared spectroscopy. Water bands at different frequencies denote the presence of non bonded to tetracoordinated waters [35, 39].

A closer analysis by molecular simulation allows to explain that the broad bands can be ascribed to subpopulations of water molecule in four of the five groups described (**Figure 9**) [30].

Hence, in a population of molecules with three hydrogen bonds, the following combinations can be possible: www, ppp, ccc, wwp, wwc, wpp, wcc, wcp, ppw, ppc, etc. This means that the water distribution and the energy of the surface are extremely heterogeneous and hence a great versatility in reaction can be expected, even more if we consider that these combinations have a mean life time. Hence, the sites can be modified by fluctuations and these can in turn affect the presence of solute from the media and the lateral interaction at different lateral pressure of the bilayer. In this regard, the presence of proteins can alter this picture.

For a better understanding of the influence of the lateral pressure on the membrane properties, we first deal with thermodynamic aspects of lipid monolayer. We will connect these properties with bilayer in the last part of the next section.

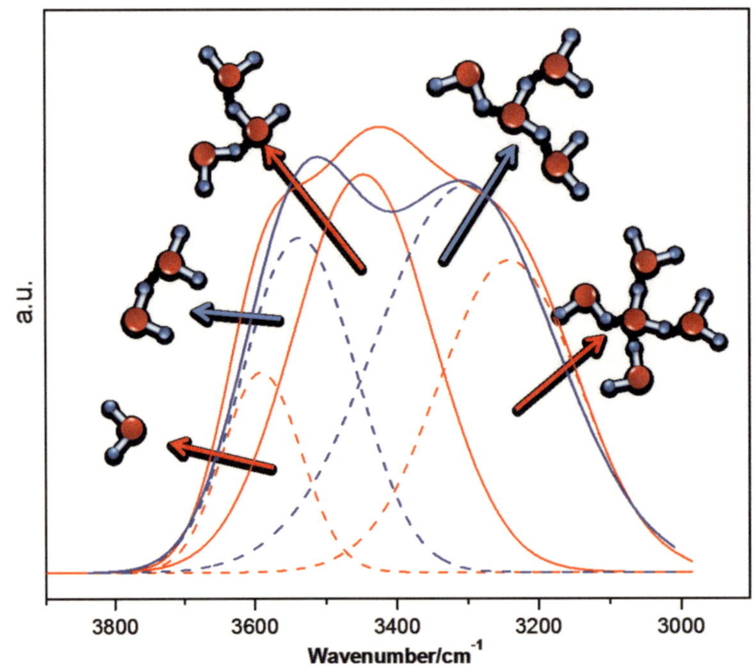

Figure 8.
Water bands in the adjacencies of lipid membranes in the gel (red line) and in the liquid crystalline state (blue line). Dotted lines denote the populations of water with none, one, two, three and four H-bonds. Central band at approximately 3500 cm^{-1} (red solid line) corresponds to bulk water without lipids.

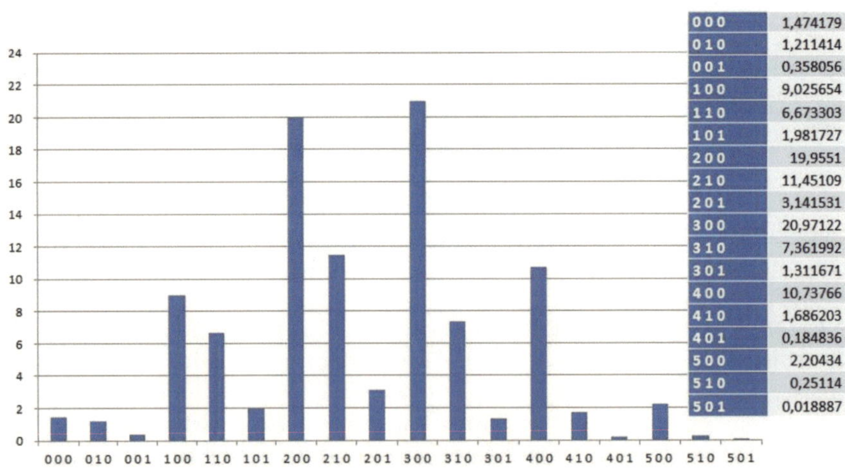

Figure 9.
Distributions of subpopulations of water forming none, one, two, three and four H bonds. Numbers correspond to water (w), carbonyl (C) and phosphate (P). Thus, 2 1 0 means two H-bonds with water, one with carbonyl, none with phosphates.

4. Stability and membrane response: a dynamic picture

The thermodynamic stabilization of the lipids in water to form bilayers as a consequence of the hydrophobic interaction between acyl chains has been discussed in several previous paper and will not be analyzed here [40–42].

Instead, the thermodynamic aspects of the bilayer as a reactive surface considering its *mechanical properties* will specifically be discussed. This approach is based on the proposals presented earlier by Damoradan et al., Cevc, Disalvo et al. [37, 38, 43–45]. In terms of interfacial properties, it is important to consider lipid interphases results that may give complementary information to present a more rigorous picture in terms of thermodynamic response. This implies to focus on the interfacial properties of the lipid arrangements both in bilayers and monolayers in which water is a component of the structure. For this reason, some general properties of lipid monolayers will be discussed to relate them in response with lipid bilayers of liposomes and vesicles to have a general formalism that can explain the behavior of both systems.

The main property inherent to membrane stabilization as a bilayer or as a monolayer is the *surface tension of water*. In the formation of bilayers, the surface tension between the acyl chain and water as described in **Figure 2** is the main force driven the *hydrophobic interaction*. The increase in entropy is expressed by the decrease in the surface tension, that is, of the surface free energy. Hence the process is spontaneous.

When lipids spread on the air-water surface (**Figure 10**), monolayers are formed spontaneously by the same reason: decrease of the surface tension with respect to that of pure water (γ^0) [46]. This is generally expressed as the *surface pressure* (Π) which is given by:

$$\Pi = \gamma^0 - \gamma \tag{1}$$

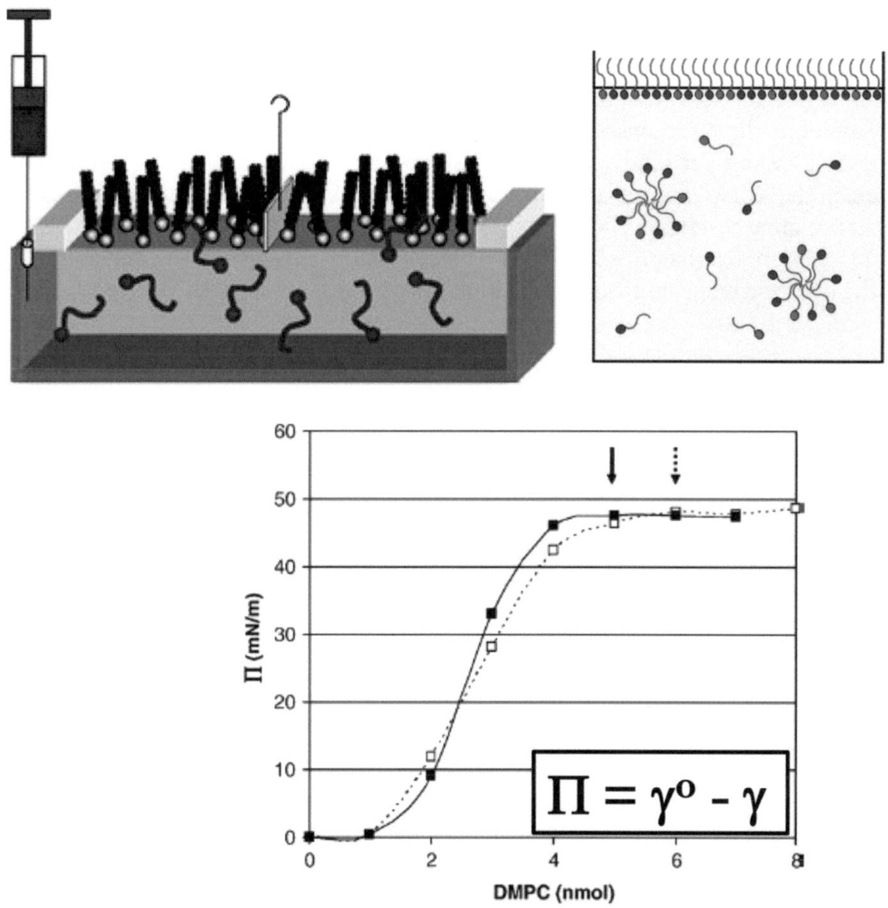

Figure 10.
The addition of lipids to the air water surface in a Langmuir though at constant area increases the surface pressure (Π) by decreasing the water surface tension (γ⁰) until a limit value is reached. From this point, the area per lipid molecule can be calculated.

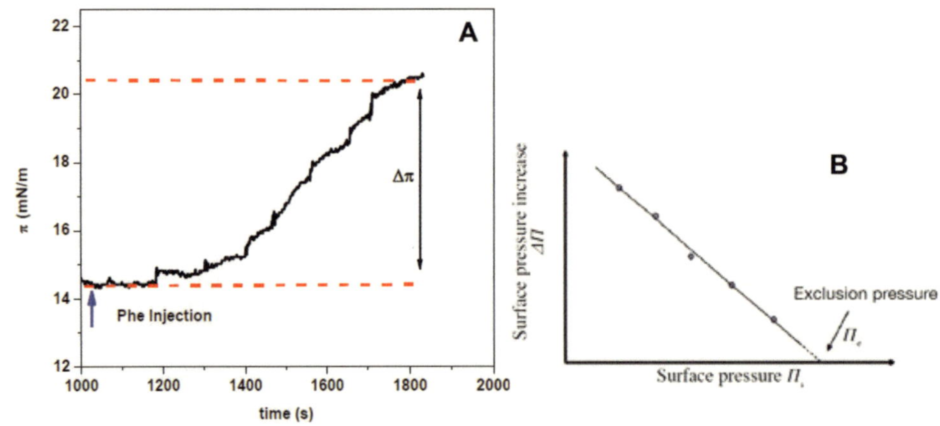

Figure 11.
(A) Kinetic of surface pressure changes after the injection of the effector to the subphase. The extent of the change reflects the affinity of the effector by the monolayer, (B) surface pressure increase versus initial pressure plot used for determining the exclusion pressure of the lipid monolayer.

Once the monolayer is stabilized at different lateral pressures (i. e. different surface excess of lipids), the response to perturbations promoted by solutes in the aqueous subphase can be tested (**Figure 11**).

As observed in **Figure 12**, *the exclusion pressure or cut off*, that is, the surface pressure at which no further effect on the monolayer is observed depends on the head group region for a given acyl chain length. DMPE (mean hydration ratio, 7 water molecules per lipid) shows a lower exclusion pressure (cut off), than DMPC (water ratio 22–24 water per lipid), although the slope of the curves remains constant. Thus, in the whole range of pressure, PE monolayers are less reactive than those of PC in the same conditions [47].

In addition, the magnitude of the surface pressure increase is also dependent on the hydrocarbon composition for similar head group regions. In **Figure 13**, it is

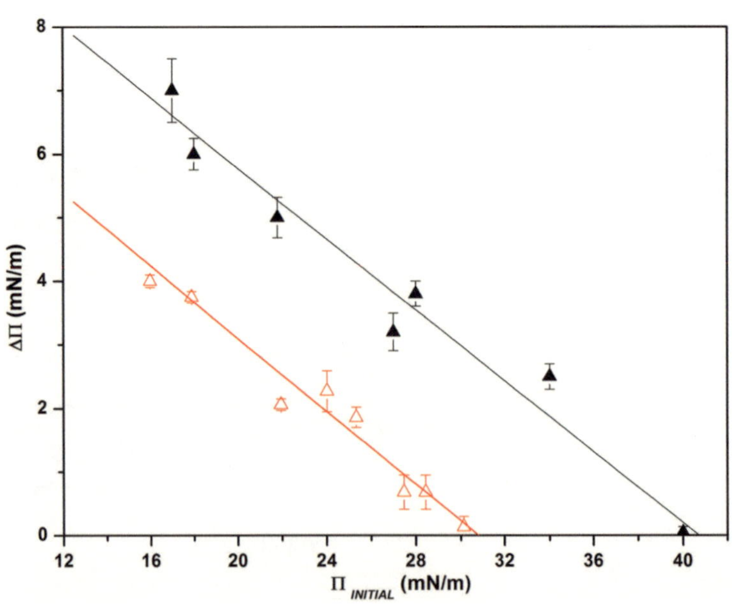

Figure 12.
The effect of a soluble protease on the surface pressure of pure DMPC (black full triangles) and DMPE (red empty triangles).

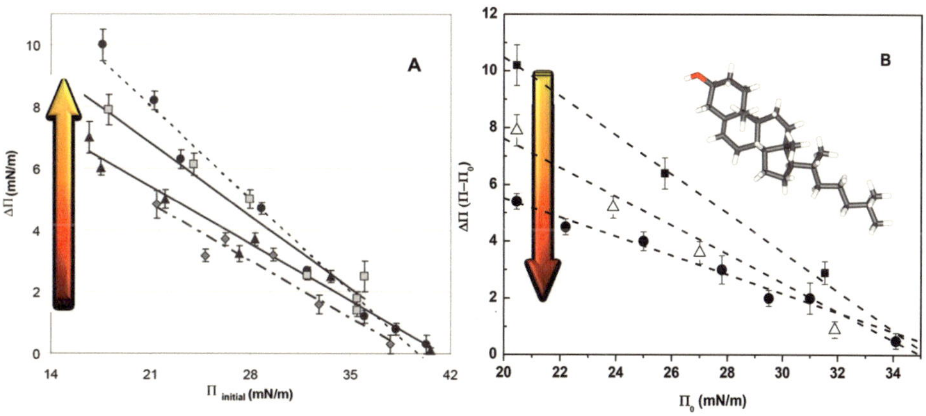

Figure 13.
Effect on the magnitude of the surface pressure response to a soluble protease of (A) unsaturation of the acyl chains in phosphatidylcholines, (B) inclusion of cholesterol in phosphatidylcholine monolayers.

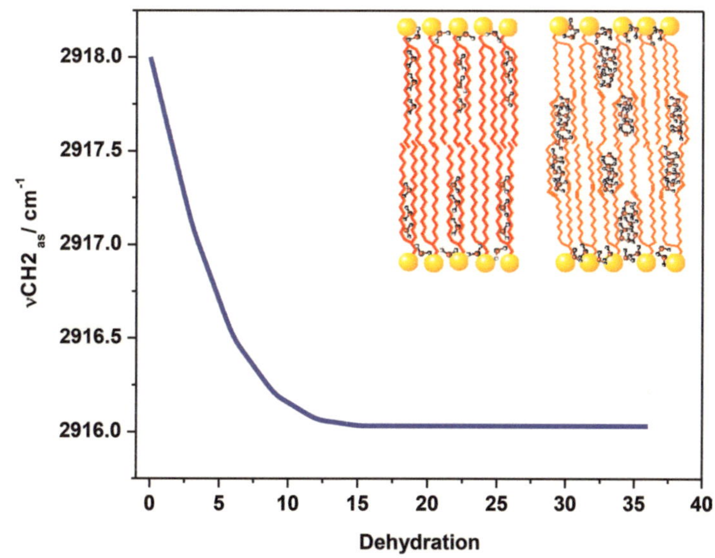

Figure 14.
Dehydration produces a decrease in confined water between acyl chains.

observed that the response at a given pressure increases with the unsaturation or ramification of the acyl chains and decreases with the presence of cholesterol [47, 48].

The surface pressures below the critical cut off (Π_c) are those states of the membrane interphase with propensity to react when chased by an effector in the subphase. These states correspond to different water/lipid ratio and hence different surface tension of water in the lipid matrix. The decrease in surface pressure $\Pi_c - \Pi$ is related to the increase of water beyond the hydration water (confined water) which part of it is distributed along the hydrocarbon chains. This can be followed by frequency changes in the methylene region by FTIR spectroscopy (**Figure 14**) [49, 50].

5. Bilayer barrier properties: permeation and controlled release

The studies in lipid monolayers allow to conclude that the response of the membrane to stimuli is due to the excess of surface tension due to the increase of water beyond the hydration shell, mostly in the acyl chain region. This excess is produced in monolayers by increasing the area per lipid, that is, reducing the surface pressure. In terms of bilayers, the expansion can be achieved by submitting liposomes to a hypotonic swelling. In this condition, membrane expands and liposomes become leaky of solute trapped in the liposome interior in isotonic conditions. Hence, *the modulation of the mechanical properties of the bilayer* by lipid composition, for instance, by including different cholesterol ratios, would allow to have a controlled release. The plots in **Figure 13** help to understand this response. At a given surface pressure, the response decreases with the increase in cholesterol, which can be related to an increase in the membrane rigidity and lower water ratio [16, 51].

Thus, *permeability* (i.e. the ratio between trapped and released solute) depends on the extent of water in the acyl region and in the second shell of phosphate groups. An important derivation of these results is that the properties of permeability can no longer be described by a model that assumes that the membrane is composed of three slabs: *a low dielectric region* (the hydrocarbon chains) and *the two aqueous interphases*, where the polar groups are located (**Figure 7**). This view predicts that solute permeating the membrane would be those that can partition in a non-polar media [52]. However, *polar amino acids and hydrophilic compounds* such as glycerol, erythritol and urea permeate easily the lipid bilayer. In addition, when swelling is taken place, ions and large molecules such as sucrose trapped in liposomes in isotonic conditions can permeate when they osmotically swell [53, 54].

Therefore, solute permeation involved in the release of trapped solutes is a more complex phenomenon than a simple *partition process*. In terms of water participation at the membrane interphase and in the acyl chain region, the mechanism would involve water reorganization and hence structural changes. To put this complexity into relevance, the interaction of polyarginines (Arg 9) with bilayers of lipids differing in hydration such as phosphatidylcholines and phosphatidylethanolamines will be discussed. In **Figure 15**, the adsorption isotherm of Arg 9 on liposomes composed by PC and PE are shown. Both present typical curves of saturation, denoting a limited number of sites with the affinity to bind Arg 9 are present in each surface [55]. They can be described by Eq. (2):

$$\theta = \frac{\Delta\zeta}{\Delta\zeta \max} = \frac{[Arg]^n}{K + [Arg]^n} \qquad (2)$$

where θ is the degree of coverage of the liposome by Arg 9, *K* the dissociation constant and *n* is a stoichiometric coefficient of the binding. In the present case, θ has been calculated by the change in the surface potential at each Arg 9 concentration measured by *electrophoretic mobility* from plots as shown in part A of **Figure 15**.

The fitting of the curves of part B gives an affinity constant for DMPC at 18°C, $K = 0.54 \times 10^3 \pm 44\ M^{-1}$ and a value $n = 1$. For DMPE at 18°C, $K = 2 \times 10^3 \pm 189\ M^{-1}$ and $n = 0.74 \pm 0.06$. According to these results, the affinity of Arg 9 is higher in DMPC than in DMPE. In addition, the interaction of Arg 9 with DMPC is well described by a *Langmuir isotherm*, since $n = 1$. In contrast, the value of n is different from 1 for Arg 9 in DMPE suggesting different mechanisms of interaction. The strong difference in affinity between PE and PC membrane can be explained recalling the monolayer results. The cut off described in **Figure 12** is much lower in PE than in PC, that is, at pressure above 32 mN/m at which PE is not affected, PC monolayer is still *reactive*. This can be explained by the strong interaction between the polar head groups of the PE molecules that restrict the water excess and hence decreasing responsiveness.

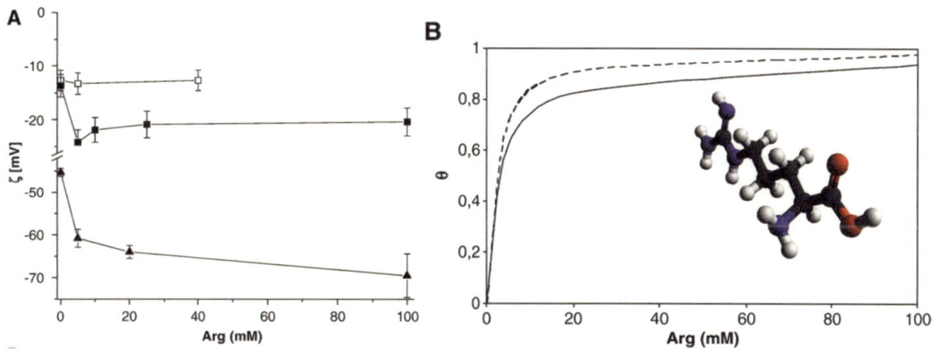

Figure 15.
Adsorption of Arg 9 on DMPC and DMPE liposomes (A) change in zeta potential (ζ) of DMPC (▲) and DMPE (■) liposomes with addition of increasing concentrations of Arg 9. θ is calculated by $\zeta - \zeta^{o}/\zeta^{\infty} - \zeta^{o}$, where ζ is the value at a given Arg 9 concentration, ζ^{o} corresponds to liposomes in the absence of Arg 9 and ζ^{∞} the value at saturation. (B) Degree of coverage as calculated from part A versus Arg 9 concentration.

6. Topological effects of osmotic shrinkage. Defects in packing

As described above, liposomes and vesicles, as well as cells, respond to *osmometers* [54]. That is, their volumes increase or decrease as a consequence of the entrance or exit of water driven by osmotic gradients. In the previous section, we show that in swelling conditions, bilayer expansion produces changes in its permeability, that is, osmosis does not only affect the liposome volume but also the membrane density. Now we will see that it also affects the surface properties.

In hypertonic media, liposomes shrink, that is, they *expulse the free water in the internal volume*. However, the volume decrease has a limit due to the repulsive forces put in evidence at short distance as described in **Figure 6**. The overcoming of this repulsive barrier derives in adhesion or fusion of the surfaces and implies *water elimination from the interphase*. The compression process due to osmotic shrinkage in bilayers can be compared with the decrease in area in monolayers. In this system, at high pressures (i.e. low areas), monolayer collapses and lipids are lost. In bilayers, the compression has different consequences due to the impossibility to extrude lipids from the bilayer.

Bilayers are soft and dynamic material [56, 57], and as a result, they can bend and deform in response to different stimuli such as dehydration and molecules that may compete with water. So, compression induced by osmotic shrinkage may result in topological changes giving place to regions (domains) of high curvature (invaginations or vesiculations) [58–60]. Thus, bilayer deformation can allow transient defects, when exposed to osmotic gradients (**Figure 16**). These curvature domains have important consequences in the interfacial behavior in the absorption or penetration of peptides composed by different types of amino acids. Many simulation studies showed that defect formation can determine the free energies of many membrane processes [61, 62].

When DPPC LUVs were subjected to hypertonic stress, defects caused by dehydration have more affinity for lytic compounds and amino acids such as phenylalanine [63–66]. The presence of defects in the membrane packing determines the binding and stabilization of peptides containing Arg and Phe motifs [67–69]

The disruption of the water network around the phenyl group and the membrane defect has been invoked to explain the *negative free energy* of the formation a PC-Phe (phosphocholine-phenylalanine) complex in the presence of water. An important observation was that a dipole potential decrease was produced in this interaction which was explained by the orientation of the carboxylate opposing to the CO of the lipids [66, 70].

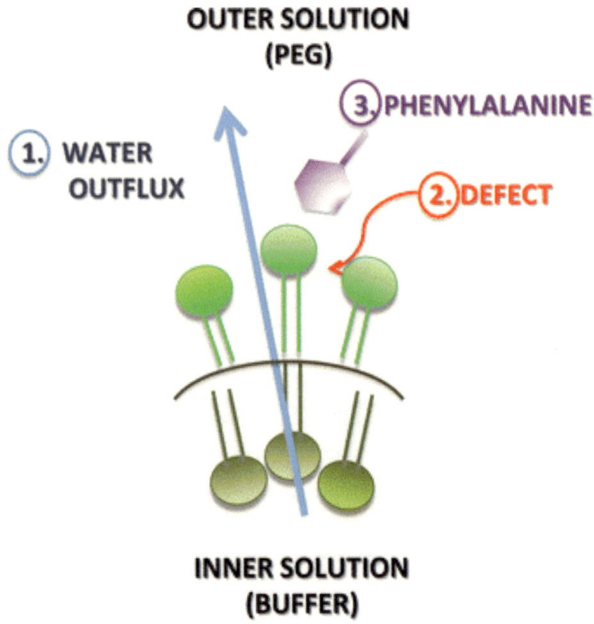

Figure 16.
Defects induced by osmotic shrinkage enhance Phe insertion into lipid bilayers.

As described in **Figure 9**, carbonyl groups are one of the hydration centers in which interfacial water is distributed. The formation of high curvature surfaces is related to changes in the *CO arrangements* [39] (**Figure 17**).

The curvature domains, in fact, increases the bilayer free energy surface by exposing hydrophobic region and carbonyl configuration as observed in part B of **Figure 17**. The stabilization of peptides or amino acid is the result of the decrease of free energy at expense of the bending modulus, that is, the energy cost of topological changes [67, 71, 72]. This last quantity is an important parameter that governs a membrane's tendency for defect formation. Many kinds of defects may be formed, including vesicle budding and fusion, depending on the presence of non-bilayer lipid phases, cholesterol and interactions between lipid bilayers and other biomolecules [19, 73].

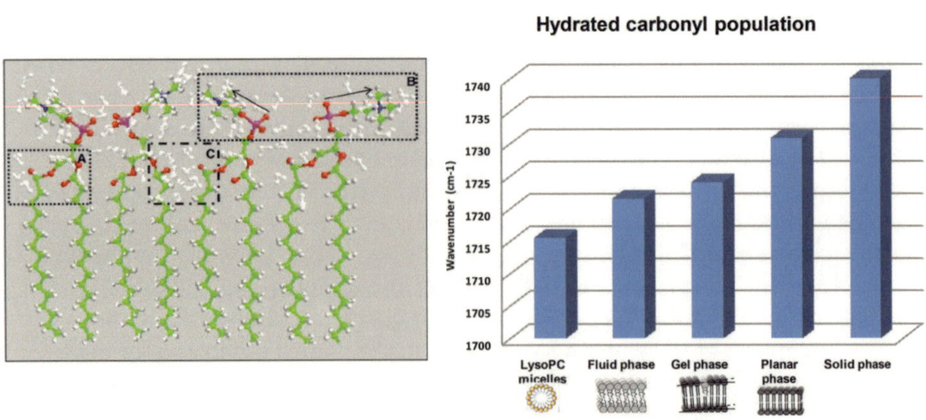

Figure 17.
Orientation of CO groups at the bilayer interphase in relation to curvature.

DOI: http://dx.doi.org/10.5772/intechopen.85811

7. Concluding remarks: design and modulation

The lipid matrix in cell membranes presents a variety of lipids. As shown in **Figure 18**, all of them present a diacyl glycerol structure with acyl chains differing in saturation and length, a phosphate group in position 3 linked to a proton, choline, ethanolamine, glycerol, serine or inositol residues. It is difficult to accept that within the economicity principle of biology, this wide variation does not play a role in *functional properties of the membrane.*

It must be noticed that all lipids may be aligned along a plane containing the glycerol backbone and the carbonyl groups. On one side of this plane, the hydrocarbon region can vary in terms of chain length, saturation and ramification. Each of these variations may give a complex matrix in which water cannot be excluded and constitute a media of a wide range of dielectric properties. On the other, polar groups protrude into the aqueous phase at different magnitudes. Importantly, also these groups are able in different extents to form hydrogen bonds between its lipid neighbors and with water. *In conclusion, it is an oversimplification to reduce the bilayer structure in which lipids organize facing the polar groups to water and segregating the nonpolar chains*.

The lack of details in these regions in relation to their emergent properties is a major limitation in the design of biomimetic systems that, as liposomes and vesicles, may be used for biotechnological and medical purposes. As a consequence, literature is saturated with works in which only "try and error" strategies are employed.

In this chapter, we have briefly discussed that:

- the interface is a bidimensional solution of hydrated polar groups.
- excess water beyond the hydration shell has solvent properties for additives in the bulk water phase and confers free energy that is excess for binding of amino acids and peptides.
- Dissolution in the water membrane phase changes the water activity (a_w) and affects the surface pressure.

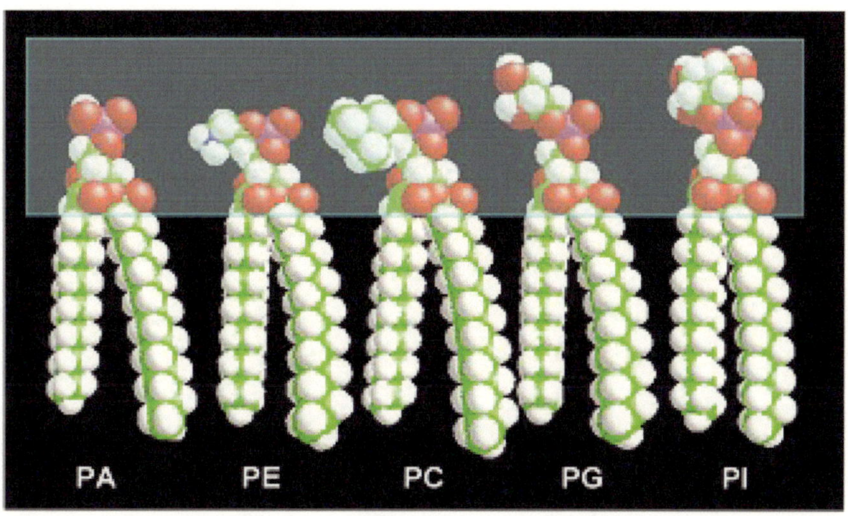

Figure 18.
Protrusion of the groups esterified to the phosphate in the surface of the membrane promotes different water organization.

These studies indicate a deeper understanding of the role of lipid bilayers in cellular biology and support the development of future lipid-based biotechnology that should necessarily include the role of water as a membrane dynamic component.

Author details

Anibal Disalvo* and Maria de los Angeles Frias
Applied Biophysics and Food Research Center, Centro de Investigaciones en Biofísica Aplicada y Alimentos, CIBAAL, National University of Santiago del Estero and CONICET, Santiago del Estero, Argentina

*Address all correspondence to: disalvoanibal@yahoo.com.ar

References

[1] Robertson JD. Membrane structure. The Journal of Cell Biology. 1981;**91**:189

[2] Bernardino de la Serna J, Schütz GJ, Eggeling C, Cebecauer M. There is no simple model of the plasma membrane organization. Frontiers in Cell and Development Biology. 2016;**4**:106

[3] Benson A. On the orientation of lipids in chloroplast and cell membranes. Journal of the American Oil Chemists Society. 1966;**43**:265-270

[4] Stoeckenius W, Engelman DM. Current models for the structure of biological membranes. The Journal of Cell Biology. 1969;**42**:613

[5] Escribá PV, Nicolson GL. Membrane structure and function: Relevance of lipid and protein structures in cellular physiology, pathology and therapy. Biochimica et Biophysica Acta. 2014;**1838**:1449

[6] Goñi FM. The basic structure and dynamics of cell membranes: An update of the Singer-Nicolson model. Biochimica et Biophysica Acta (BBA)-Biomembranes. 2014;**1838**:1467-1476

[7] Bagatolli LA, Ipsen JH, Simonsen AC, Mouritsen OG. An outlook on organization of lipids in membranes: Searching for a realistic connection with the organization of biological membranes. Progress in Lipid Research. 2010;**49**:378-389

[8] Bangham A, Hill M, Miller N. Preparation and Use of Liposomes as Models of Biological Membranes. New York: Springer; 1974

[9] Bangham A. Lipid bilayers and biomembranes. Annual Review of Biochemistry. 1972;**41**:753-776

[10] Wender PA, Kreider E, Pelkey ET, Rothbard J, VanDeusen CL. Dendrimeric molecular transporters: Synthesis and evaluation of tunable polyguanidino dendrimers that facilitate cellular uptake. Organic Letters. 2005;**7**:4815-4818

[11] Torchilin VP, Rammohan R, Weissig V, Levchenko TS. TAT peptide on the surface of liposomes affords their efficient intracellular delivery even at low temperature and in the presence of metabolic inhibitors. Proceedings of the National Academy of Sciences. 2001;**98**:8786-8791

[12] Gregoriadis G. Liposome Technology. Boca Raton, FL: CRC Press; 1992

[13] Luzzati V, Tardieu A, Taupin D. A pattern-recognition approach to the phase problem: Application to the X-ray diffraction study of biological membranes and model systems. Journal of Molecular Biology. 1972;**64**:269-286

[14] Quinn PJ, Wolf C. An X-ray diffraction study of model membrane raft structures. The FEBS Journal. 2010;**277**:4685-4698

[15] Caffrey M. The study of lipid phase transition kinetics by time-resolved x-ray diffraction. Annual Review of Biophysics and Biophysical Chemistry. 1989;**18**:159-186

[16] Ladbrooke B, Williams RM, Chapman D. Studies on lecithin-cholesterol-water interactions by differential scanning calorimetry and X-ray diffraction. Biochimica et Biophysica Acta (BBA)-Biomembranes. 1968;**150**:333-340

[17] Chapman D. Phase transitions and fluidity characteristics of lipids and cell membranes. Quarterly Reviews of Biophysics. 1975;**8**:185-235

[18] Marsh D. Handbook of Lipid Bilayers. Boca Raton, FL: CRC Press; 2013

[19] Gruner SM. Intrinsic curvature hypothesis for biomembrane lipid composition: A role for nonbilayer lipids. Proceedings of the National Academy of Sciences. 1985;**82**:3665-3669

[20] Cullis PT, Kruijff BD. Lipid polymorphism and the functional roles of lipids in biological membranes. Biochimica et Biophysica Acta (BBA)-Reviews on Biomembranes. 1979;**559**:399-420

[21] Tristram-Nagle S, Nagle JF. Lipid bilayers: Thermodynamics, structure, fluctuations, and interactions. Chemistry and Physics of Lipids. 2004;**127**:3-14

[22] Nickels JD, Katsaras J. Water and lipid bilayers. In: Membrane Hydration. Switzerland: Springer; 2015. pp. 45-67

[23] Disalvo EA. Permeability of water and polar solutes in lipid bilayers. Advances in Colloid and Interface Science. 1988;**29**:141-170

[24] Disalvo EA, Simon SA. Permeability and Stability of Lipid Bilayers. Boca Raton, FL: CRC Press; 1995

[25] MacDonald R, Simon S. Lipid monolayer states and their relationships to bilayers. Proceedings of the National Academy of Sciences. 1987;**84**:4089-4093

[26] Luzardo MdC, Peltzer G, Disalvo E. Surface potential of lipid interfaces formed by mixtures of phosphatidylcholine of different chain lengths. Langmuir. 1998;**14**:5858-5862

[27] Rand R, Parsegian V. Hydration forces between phospholipid bilayers. Biochimica et Biophysica Acta (BBA)-Reviews on Biomembranes. 1989;**988**:351-376

[28] Nagle JF, Tristram-Nagle S. Structure of lipid bilayers. Biochimica et Biophysica Acta. 2000;**1469**:159-195

[29] Disalvo EA. Membrane Hydration: The Role of Water in the Structure and Function of Biological Membranes. Switzerland: Springer; 2015

[30] Alarcón LM, de los Angeles Frías M, Morini MA, Sierra MB, Appignanesi GA, Disalvo EA. Water populations in restricted environments of lipid membrane interphases. The European Physical Journal E. 2016;**39**:94

[31] Arrondo J, Goni F, Macarulla J. Infrared spectroscopy of phosphatidylcholines in aqueous suspension a study of the phosphate group vibrations. Biochimica et Biophysica Acta (BBA)-Lipids and Lipid Metabolism. 1984;**794**:165-168

[32] Goni FM, Arrondo JL. A study of phospholipid phosphate groups in model membranes by Fourier transform infrared spectroscopy. Faraday Discussions of the Chemical Society. 1986;**81**:117-126

[33] Wiener M, Suter R, Nagle J. Structure of the fully hydrated gel phase of dipalmitoylphosphatidylcholine. Biophysical Journal. 1989;**55**:315-325

[34] Arsov Z. Long-range lipid-water interaction as observed by ATR-FTIR spectroscopy. In: Membrane Hydration. Switzerland: Springer; 2015. pp. 127-159

[35] Disalvo EA, Frias MA. Water state and carbonyl distribution populations in confined regions of lipid bilayers observed by FTIR spectroscopy. Langmuir. 2013;**29**:6969-6974

[36] Lewis RN, McElhaney RN. The structure and organization of phospholipid bilayers as revealed by infrared spectroscopy. Chemistry and Physics of Lipids. 1998;**96**:9-21

[37] Damodaran S. Water activity at interfaces and its role in regulation of interfacial enzymes: A hypothesis.

Colloids and Surfaces B: Biointerfaces. 1998;**11**:231-237

[38] Disalvo EA, Hollmann A, Semorile L, Martini MF. Evaluation of the Defay-Prigogine model for the membrane interphase in relation to biological response in membrane-protein interactions. Biochimica et Biophysica Acta. 2013;**1828**:1834-1839

[39] Frias Mde L, Disalvo EA. Configuration of carbonyl groups at the lipid interphases of different topological arrangements of lipid dispersions. Langmuir. 2009;**25**:8187-8191

[40] Boggs JM. Lipid intermolecular hydrogen bonding: Influence on structural organization and membrane function. Biochimica et Biophysica Acta (BBA)-Reviews on Biomembranes. 1987;**906**:353-404

[41] McMullen TP, McElhaney RN. Differential scanning calorimetric studies of the interaction of cholesterol with distearoyl and dielaidoyl molecular species of phosphatidylcholine, phosphatidylethanolamine, and phosphatidylserine. Biochemistry. 1997;**36**:4979-4986

[42] Israelachvili J, Marčelja S, Horn RG. Physical principles of membrane organization. Quarterly Reviews of Biophysics. 1980;**13**:121-200

[43] Disalvo EA. Membrane hydration: A hint to a new model for biomembranes. Sub-Cellular Biochemistry. 2015;**71**:1-16

[44] Cevc G. How membrane chain melting properties are regulated by the polar surface of the lipid bilayer. Biochemistry. 1987;**26**:6305-6310

[45] Sparr E, Wennerström H. Responding phospholipid membranes—Interplay between hydration and permeability. Biophysical Journal. 2001;**81**:1014-1028

[46] Lairion F, Disalvo EA. Effect of dipole potential variations on the surface charge potential of lipid membranes. The Journal of Physical Chemistry. B. 2009;**113**:1607-1614

[47] Martini M, Disalvo E. Superficially active water in lipid membranes and its influence on the interaction of an aqueous soluble protease. Biochimica et Biophysica Acta (BBA)-Biomembranes. 2007;**1768**:2541-2548

[48] Martini MF, Disalvo EA. Effect of polar head groups on the activity of aspartyl protease adsorbed to lipid membranes. Chemistry and Physics of Lipids. 2003;**122**:177-183

[49] Disalvo EA, Bouchet AM, Frias MA. Connected and isolated CH_2 populations in acyl chains and its relation to pockets of confined water in lipid membranes as observed by FTIR spectrometry. Biochimica et Biophysica Acta. 2013;**1828**:1683-1689

[50] Pinto OA, Bouchet AM, Frias MA, Disalvo EA. Microthermodynamic interpretation of fluid states from FTIR measurements in lipid membranes: A Monte Carlo study. The Journal of Physical Chemistry. B. 2014;**118**:10436-10443

[51] Gong K, Feng S-S, Go ML, Soew PH. Effects of pH on the stability and compressibility of DPPC/cholesterol monolayers at the air-water interface. Colloids and Surfaces A: Physicochemical and Engineering Aspects. 2002;**207**:113-125

[52] Wimley WC, White SH. Membrane partitioning: Distinguishing bilayer effects from the hydrophobic effect. Biochemistry. 1993;**32**:6307-6312

[53] Disalvo EA, Campos AM, Abuin E, Lissi EA. Surface changes induced by osmotic shrinkage on large unilamellar vesicles. Chemistry and Physics of Lipids. 1996;**84**:35-45

[54] Kinnunen PK. Lipid bilayers as osmotic response elements. Cellular Physiology and Biochemistry. 2000;**10**:243-250

[55] Disalvo E, Bouchet A. Electrophoretic mobility and zeta potential of liposomes due to arginine and polyarginine adsorption. Colloids and Surfaces A: Physicochemical and Engineering Aspects. 2014;**440**:170-174

[56] Evans E, Skalak R. Mechanics and thermodynamics of biomembranes: Part 2. CRC Critical Reviews in Bioengineering. 1979;**3**:331

[57] Marsh D. Elastic curvature constants of lipid monolayers and bilayers. Chemistry and Physics of Lipids. 2006;**144**:146-159

[58] Evans EA, Parsegian VA. Energetics of membrane deformation and adhesion in cell and vesicle aggregation. Annals of the New York Academy of Sciences. 1983;**416**:13-33

[59] Epand R, Fuller N, Rand R. Role of the position of unsaturation on the phase behavior and intrinsic curvature of phosphatidylethanolamines. Biophysical Journal. 1996;**71**:1806-1810

[60] McMahon HT, Gallop JL. Membrane curvature and mechanisms of dynamic cell membrane remodelling. Nature. 2005;**438**:590

[61] Baumgart T, Capraro BR, Zhu C, Das SL. Thermodynamics and mechanics of membrane curvature generation and sensing by proteins and lipids. Annual Review of Physical Chemistry. 2011;**62**:483-506

[62] MacCallum JL, Bennett WD, Tieleman DP. Distribution of amino acids in a lipid bilayer from computer simulations. Biophysical Journal. 2008;**94**:3393-3404

[63] Senisterra GA, Disalvo EA, Gagliardino JJ. Osmotic dependence of the lysophosphatidylcholine lytic action on liposomes in the gel state. Biochimica et Biophysica Acta. 1988;**941**:264-270

[64] Cutro A, Disalvo E. Phenylalanine blocks defects induced in gel lipid membranes by osmotic stress. The Journal of Physical Chemistry B. 2015;**119**:10060-10065

[65] Frias MA, Nicastro A, Casado NM, Gennaro AM, Diaz SB, Disalvo EA. Arbutin blocks defects in the ripple phase of DMPC bilayers by changing carbonyl organization. Chemistry and Physics of Lipids. 2007;**147**:22-29

[66] Rosa AS, Cutro AC, Frias MA, Disalvo EA. Interaction of phenylalanine with DPPC model membranes: More than a hydrophobic interaction. The Journal of Physical Chemistry. B. 2015;**119**:15844-15847

[67] Nuscher B, Kamp F, Mehnert T, Odoy S, Haass C, Kahle PJ, et al. α-Synuclein has a high affinity for packing defects in a bilayer membrane a thermodynamics study. Journal of Biological Chemistry. 2004;**279**:21966-21975

[68] Mitchell DJ, Steinman L, Kim D, Fathman C, Rothbard J. Polyarginine enters cells more efficiently than other polycationic homopolymers. The Journal of Peptide Research. 2000;**56**:318-325

[69] Griffith EC, Perkins RJ, Telesford D-M, Adams EM, Cwiklik L, Allen HC, et al. Interaction of L-phenylalanine with a phospholipid monolayer at the water-air interface. The Journal of Physical Chemistry B. 2014;**119**:9038-9048

[70] Cutro AC, Hollmann A, Cejas J, Maturana P, Disalvo EA, Frias MA. Phenylalanine interaction with lipid monolayers at different pHs. Colloids and Surfaces. B, Biointerfaces. 2015;**135**:504-509

[71] Ziegler A, Li Blatter X, Seelig A, Seelig J. Protein transduction domains of HIV-1 and SIV TAT interact with charged lipid vesicles. Binding mechanism and thermodynamic analysis. Biochemistry. 2003;**42**:9185-9194

[72] Gonçalves E, Kitas E, Seelig J. Binding of oligoarginine to membrane lipids and heparan sulfate: Structural and thermodynamic characterization of a cell-penetrating peptide. Biochemistry. 2005;**44**:2692-2702

[73] Marsh D. Intrinsic curvature in normal and inverted lipid structures and in membranes. Biophysical Journal. 1996;**70**:2248-2255

Chapter 3

General Perception of Liposomes: Formation, Manufacturing and Applications

Christian Isalomboto Nkanga, Alain Murhimalika Bapolisi, Nnamdi Ikemefuna Okafor and Rui Werner Maçedo Krause

Abstract

Liposomes are currently part of the most reputed carriers for various molecular species, from small and simple to large and complex molecules. Since their discovery, liposomes have been subject to extensive evolution, in terms of composition, manufacturing and applications, which led to several openings in both basic and applied life sciences. However, most of the advances in liposome research have been more devoted to launching new developments than improving the existing technology for potential implementation. For instance, the evolution of the conventional lipid hydration methods to novel microfluidic technologies has permitted upscale production, but with increase in manufacturing cost and persistent use of organic solvents. This chapter intends to present general concepts in liposome technology, highlighting some long-standing bottlenecks that remain challenging to the preparation, characterization and applications of liposomal systems. This would enhance the understanding of the gaps in the field and, hence, provide directions for future research and developments.

Keywords: phospholipids, soybean lecithin, liposome composition, manufacturing methods, characterization techniques

1. Introduction

Liposomes are artificial lipid-based bilayered vesicles. They were firstly discovered and described in 1965 by Bangham et al. [1] as swollen phospholipid systems, namely Banghasomes. A few years later, the structural description of liposomes was unveiled as small devices made of one or more closed phospholipid bilayers. Due to the diversity of particle sizes, from 20 nanometers to several micrometers, liposomal vesicles are considered as either nanoparticles or microparticles endowed with the ability to encapsulate materials of various nature and polarity [2, 3].

Up until now, liposomes have shown huge promise as potential vehicles for biologically active compounds in cosmetic and pharmaceutical industries. These applications have been extended to food and farm industries, where unstable substances such as antioxidants, flavors and antimicrobials have been explored for liposomal encapsulation. Across all these areas of application, liposomes have been deemed to be the most successful delivery systems due to their multiple advantages. These include high biocompatibility and biodegradability, low toxicity and poor immunogenicity, improved drug solubility and controlled distribution, as well as the ability of performing surface modifications for targeted, extended and sustained release. Currently, there are several liposomal formulations that are clinically established for the treatment of various diseases, such as cancer, fungal and viral infections; and many more have reached advanced phases of clinical trials [4, 5].

Although liposomes have shown some success in drug product development, the limitations identified in liposomal technology have remained almost stagnant over decades. The most common disadvantages of liposomes arise partly from poor stability under shelf and in vivo conditions. This is mostly due to potential lipids oxidation and hydrolysis, leakage and loss of hydrophilic cargoes, as well as particles fission and fusion. To date, some of these problems can be circumvented by playing around formulation adjuvants, such as anti-oxidants, or post-preparation processing, such as freeze-drying [4, 6, 7].

While describing broadly the current perception of liposomes, regarding production, evaluation and applications; this chapter intends to highlight the longstanding bottlenecks that remained overlooked and challenging to product development and implementation. This would increase the understanding of the gaps in the field and provide future directions to new openings for improvements in liposome technology.

2. Liposomes formation and classification

2.1 Liposomes formation

The liposomal vesicles derive from hydration of phospholipids, which are amphiphilic molecules endowed with a hydrophilic head group and two hydrophobic acyl chains (**Figure 1**). In aqueous media, phospholipid molecules self-assemble

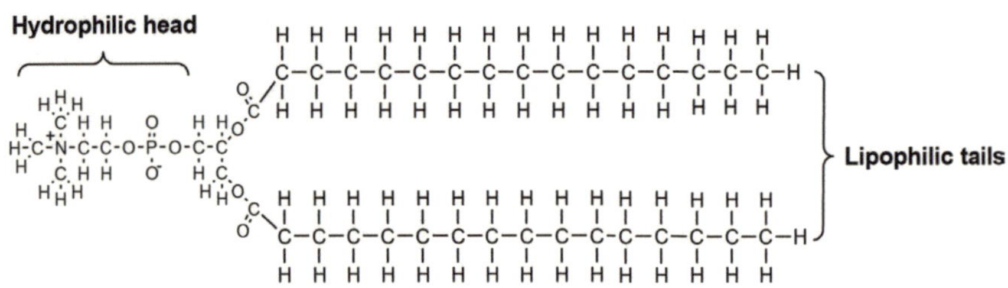

Figure 1.
Chemical structure of a representative phospholipid molecule (distearoyl phosphatidylcholine).

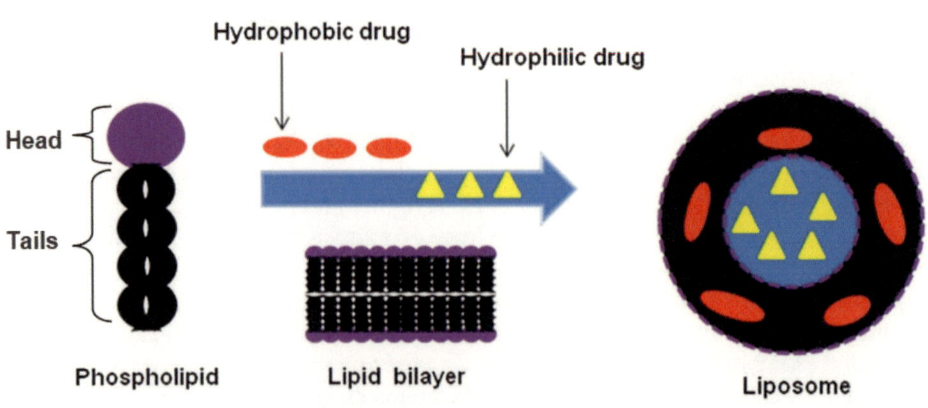

Figure 2.
Flowchart illustrating liposome formation and encapsulation of drug molecules.

Vesicle designation	Main components	Illustrative application
Emulsomes	A mixture of fats and triglycerides stabilized by high proportion of lecithin	Emulsomes loaded with Amphotericin B for the treatment of visceral leishmaniasis [9]
Enzymosomes	Complexes of lipids and enzymatic proteins	Encapsulation and delivery of superoxide dismutase for oxidative stress management [9]
Sphyngosomes	Sphingolipids containing amide and ether bonds	Sphyngosomes loaded with vincristine (Marqibo®) for lymphoblastic leukemia therapy [4]
Transfersomes	A mixture of single chain surfactant, phospholipids and ethanol (10%)	Transfersomes loaded with diclofenac for improved topical delivery/retention [10]
Ethosomes	Phospholipids and ethanol (20–40%)	Mitoxantrone-loaded ethosomes for the treatment of melanoma [11]
Pharmacosomes	Conjugate of drug and phospholipid	Pharmacosomes loaded with diclofenac for enhanced the bioavailability and reduced toxicity [9]
Virosomes	Viral glycoproteins	Virosome containing HIV-1 gp41-subunit antigens for protection against vaginal simian-HIV [12]
Aquasomes	Tin oxide, diamonds or brushite core covered with oligomeric film	PEG-lipid coated aquasomes containing interferon-α-2b for prolonged and enhanced cytotoxicity [13]
Bilosomes	Bile salts and acids (deoxycholic acid)	Bilosomes loaded with diphtheria toxoid for systemic and mucosal immunization [14]
Niosomes	Non-ionic surfactants (span and tween)	Niosomes based formulation for enhanced oral bioavailability of candesartan cilexetil [15]

Table 1.
Presentation of liposome-type systems.

into a bilayered structure. Within the bilayer, phospholipid polar groups line up to form a water-attracting surface while their lipophilic chains face each other to yield a water-free zone. On mechanical shaking or heating, phospholipid bilayers continuously enclose the dispersing aqueous medium and form a vesicular system. In this system, hydrophilic groups of phospholipids are oriented towards the inner and outer aqueous phase, while their hydrophobic tails are centered within the bilayer [2, 4]. This architecture underlines the ability of liposomes to readily encapsulate hydrophilic and hydrophobic materials inside the inner aqueous core and the lipid bilayers, respectively (as illustrated in **Figure 2**).

2.2 Liposomes classification

Depending on the particle size and number of bilayers forming the vesicles (lamellarity), liposomes can be categorized in the following classes [4, 8]:

- Small unilamellar vesicles (SUV), size range 20–100 nm;
- Large unilamellar vesicles (LUV), size >100 nm;
- Giant unilamellar vesicles (GULV), size >1000 nm;
- Oligolamellar vesicles (OLV), size range 100–1000 nm;

- Multilamellar large vesicles (MLV), with size >500 nm;
- Multivesicular vesicles, size from 1000 nm to several thousand nanometers.

Based on their composition, liposomes can be classified as conventional, long circulating, cationic, stimuli-responsive and immunoliposomes. The differences between these categories will be highlighted later when discussing composition and evolution of liposomes.

Furthermore, there are many other vesicular systems considered as part of the liposome-type vesicles. These include emulsomes, enzymosomes, sphyngosomes, transfersomes, ethosomes, pharmacosomes and virosomes, which are lipid-based liposomes analogous. The non-lipid-based liposomes analogous are aquasomes, bilosomes and niosomes [8, 9]. All the liposome-type systems are briefly presented in **Table 1**.

3. Liposomes composition and evolution

3.1 Liposomes composition

Liposomes are made of physiologically acceptable natural or synthetic phospholipids found in the lipid bilayer membranes of human cells. The most frequently used phospholipids for liposomes preparation are phosphatidylcholines (PC), phosphatidylethanolamines (PE), phosphatidylserines (PS) and phosphatidylglycerols (PG) [16]. The molecular structures of these biocompatible lipids are shown in **Figure 3** and **Table 2**.

In liposomal technology, a considerable attention is given to the phase transition temperature (Tt) of these phospholipids. The Tt corresponds to the temperature above which phospholipids exist in liquid crystalline phase. In this fluid state, hydrophobic tails of phospholipids are randomly oriented but ready to form closely continuous bilayered vesicles (liposomes). Below the Tt, phospholipids exist in gel state, where the hydrophobic tails are completely expanded and well packed, thus not able to form liposomes [4, 17].

As most of the phospholipids used for liposomes formulation have Tt close to the physiological temperature (37°C), the addition of cholesterol has been adopted as a strategy to stabilize the liposomal vesicles in physiological media. This is especially for phospholipids that can undergo phase transition and leakage at room temperature, which can lead to premature release of the liposome cargo. In fact, due to its high hydrophobicity, cholesterol was found efficient in strengthening the packing of phospholipid bilayers, reducing therefore membrane permeability. Numerous

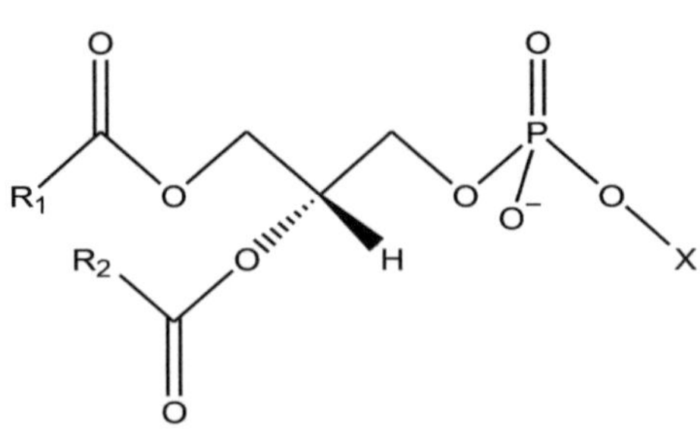

Figure 3.
Structural representation of ester glycerol-phospholipid molecules, with R1 and R2 representing the hydrocarbon chains of different fatty acids.

Phospholipids	**PC**	**PE**	**PS**	**PG**
——X Moiety	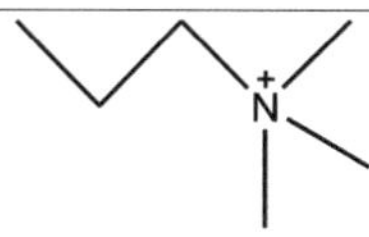		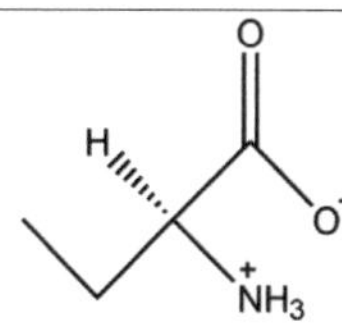	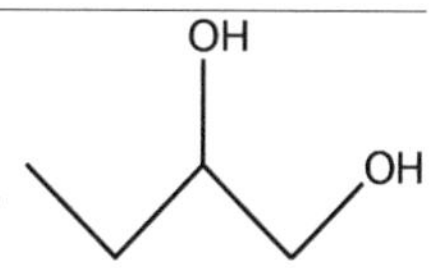

Table 2.
Description of –X moieties of different glycerol-phospholipids [16].

studies have reported the ability of cholesterol to impact liposomes properties and functionality, including encapsulation efficiency and release characteristics [18–20]. The work by Kirby et al. [21] demonstrated that increasing cholesterol content can prevent leakage and improve in vivo stability of liposome. Later, Lopez-Pinto et al. [22] observed a direct correlation between cholesterol content and liposome sizes. These observations have established cholesterol content to be a key parameter in liposome formulation.

Like cholesterol, there are many other ingredients that can affect liposomes behavior and afford the desired encapsulation or delivery profiles. Additive agents such as oleic acid and N-[1(2,3-dioleoyloxy) propyl]-N,N,N-trimethylammonium chloride (DOTAP) are useful for the preparation of negatively and positively charged liposomes, respectively. These charged liposomes offer the advantage of great liposomal stability during the storage, as charged particles repel each and reduce aggregation tendencies. While the cell internalization of positively charged liposomes (cationic liposomes) is promoted by their electrostatic interaction with cell membranes (which are negatively charged), liposomes bearing negative charges are subjective to poor cell internalization due to the corresponding repulsive forces. Cationic liposomes are used in gene therapy due to their ability to successfully encapsulate nucleic acids by electrostatic forces [4, 23].

In addition, some special lipids such as cholesteryl hemisuccinate (CHEMS) and 1,2-dioleoyl-sn-glycero-3-phosphoethanolamine (DOPE), have been widely used to prepare liposomes with pH-dependent release features. CHEMS can exhibit pH-sensitivity either alone or in the presence of other lipids. In ionized forms at basic or neutral pH, CHEMS stabilizes the lamellar form of DOPE in lipid-based vesicles. However, the protonated or molecular CHEMS formed at acidic pH promotes hexagonal phase of this lipid, which leads to the disruption of the vesicular systems and release of the encapsulated materials [24, 25]. Tocopherol hemisuccinate (THS) has also shown similar pH-responsiveness as CHEMS, due to their molecular similarity [26].

The composition of liposome appears to be a broad topic, but also very crucial for the desired product development. However, the nature and costs of the liposome components used over decades viz., particularly the synthetic or highly purified natural phospholipids, have been reported to be part of the factors affecting negatively the universal implementation and affordability of liposome technology [16, 27]. The review by Machado et al. [28] discussed the feasibility of using crude soybean and rice lecithin for liposomal encapsulation of food ingredients. The authors demonstrated these naturally occurring phospholipid mixtures could be useful for liposomes preparation regardless of the intended area of application. Our group has recently investigated liposomal encapsulation of isoniazid using crude soybean lecithin. The formulated liposomes exhibited much better encapsulation efficiency than purified soybean lecithin [29]. This study proposed crude soybean lecithin for liposomal encapsulation of drug molecules. However, the complexity of this lipid mixture might be a bottleneck for some biomedical applications, where molecular architecture of the lipid bilayer is explored to get insights into potential cell membrane permeability. The versatility of crude soybean lecithin liposomes is therefore in question, considering the wide range of areas that the liposomal systems usually cover.

3.2 Liposomes evolution

Based on the composition, liposomal systems can be considered to have evolved from conventional, long circulating, targeted and immune-liposomes to stimuli-responsive liposomes. The liposomes composed purely of phospholipids with or without cholesterol (conventional liposomes) have shown some limitations due to their uptake by the cells of the mononuclear phagocytic system (MNPS), such as macrophages that ensure liposomes clearance through phagocytosis. This biological fate makes conventional liposomes appropriate vehicles for targeted drug delivery to infected MNPS cells, like the case of alveolar macrophages where *Mycobacterium tuberculosis* resides often. However, the uptake by the MNPS cells decreases liposomes half-life and exposes to high risk of therapeutic failure when the site of interest is beyond the MNPS [2].

Extensive studies conducted in liposome technology led to identification of some astute strategies for addressing the issue of MNPS' attack viz., liposome downsizing and surface modification. In this regard, it was observed that the physiological clearance of larger liposomes (MLV) was much quicker than that of smaller liposomes (SUL), which describe long-circulating profiles with increased half-lives [30]. The stealth strategy arising from surface modification involved grafting or coating hydrophilic polymers such as polyethylene glycol (PEG) and chitosan, which prevents detection of liposomes by the MNPS cells. While stealth behavior allows liposomes to achieve much longer circulation time (hence the name "long circulating liposomes"), this strategy comes along with poor targeting efficiency due to wider distribution of liposomes in the body. From this limitation, further developments have led to introduction of targeted liposomes. These liposomes are characterized by surface decoration with glycoproteins, polysaccharides or specific receptors ligands to achieve narrowed distribution and accumulation at the site of interest [9, 31]. The observation that ligand-decorated liposomes could provide selective drug accumulation inspired further design of antibody-functionalized liposomes (immunoliposomes) as well as stimuli-responsive liposomal systems for controlled drug delivery [32]. **Figure 4** shows the trend in the development of "intelligent" liposomes for site-specific delivery; from conventional liposomes, stealth liposome, targeted liposomes, immunoliposomes to stimuli-responsive liposomes.

Stimuli-responsive liposomes are smart liposomal systems that exhibit rapid release of the cargo upon changes in some physicochemical or biochemical stimuli, such as pH, temperature, redox potentials, enzymes concentrations, ultrasound, electric or magnetic fields [33, 34]. Among these stimuli, pH change is the most promising stimulus due to the existence of multiple pH gradients in the body [35]. In common practice, formulation of pH-sensitive liposomes involves incorporation of CHEMS and DOPE. Although CHEMS-DOPE-based liposomes have shown great promise for controlled delivery [24, 25, 36], the costly status of these lipids remains a deep concern for wider development and application of liposome products. To circumvent the use of such special lipids for pH-dependent delivery, our group has considered hydrazone derivatization of isoniazid (INH), as a small hydrophilic drug model, followed by encapsulation using crude soybean lecithin for cost-effective development [37]. Being poorly water soluble, the prepared conjugates were successfully embedded within the lipid membranes and INH release experiments were conducted in different pH media. The INH-conjugate loaded crude soybean lecithin

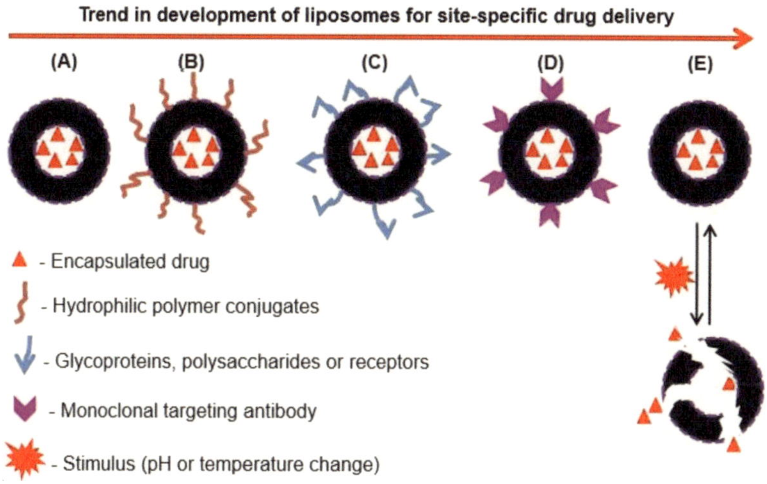

Figure 4.
Schematic representation of the trend in liposome improvements for site specific delivery. (A) Conventional liposomes, (B) stealth liposome, (C) targeted liposomes, (D) immunoliposomes and (E) stimuli-responsive liposomes [32].

liposomes have demonstrated attractive nanoparticulate and stimuli-responsive characteristics for potential low-cost site-specific liposomal delivery [38]. However, the amounts of INH-derivatives loaded were found to be almost 10-folds lesser than the loading achieved when native INH was encapsulated in crude soybean lecithin liposomes, which was in turn associated with some burst release [29]. This is probably due to the fact that INH derivatives are encapsulated in the lipid bilayer, which offers limited space for loading [39], while the native INH was trapped within the aqueous core of liposomes. These limitations underline the need for further developments in liposome technology to achieve controlled release from cost-effective liposomes, combining both use of cheap lipids and high drug loadings.

4. Liposomes preparation and characterization

4.1 Preparation methods

Liposomes can be prepared using a wide range of methods that involve combination of lipids with aqueous media, and somehow affect liposomes characteristics, such as size, lamellarity and encapsulation efficiency (EE). The recently reported methods can be categorized as conventional, which mostly involve approaches that are easy to use at laboratory scale, and novel methods that appear to be more useful for up-scale production but require some special equipment [4].

4.1.1 Conventional methods

The most commonly used methods for formulation of liposomes share the following fundamental stages: (i) lipids dissolution in organic solvents, (ii) drying of the resultant solution, (iii) hydration of dried lipid (using various aqueous media), (iv) isolation of the liposomal vesicles, and (v) quality control assays [6]. While sharing these basic stages, the conventional preparation methods gather different advantages and disadvantages that are comparatively presented in **Table 3**.
The specific technological details of these methods are separately discussed in the following paragraphs.

Method designation	Advantages	Disadvantages
Film hydration (Bangham method)	Straightforward process	Use of organic solvent and mechanical agitation, production of large particles with no control on size, poor encapsulation efficiencies of hydrophilic materials, time consuming, sterilization issue
Reverse phase evaporation	Simple design, suitable encapsulation efficiency	Not applicable to fragile cargoes, use of large quantity of organic solvent, time consuming, sterilization issue
Solvent injection	Straightforward approach	Trace of organic solvent as residue, possible nozzle blockage in ether system, time consuming, sterilization issue
Detergent removal	Simple design, homogenous product, control of particle size	Presence of organic solvent, detergent residue, time consuming, low entrapment efficiency, poor yield, sterilization required
Heating method	Simple and fast process, organic solvent free, no need for sterilization, possible up-scale production	The need for high temperature

Adapted from [40].

Table 3.
Advantages and disadvantages of conventional methods.

4.1.1.1 Film hydration

Also known Bangham method, film hydration represents the simplest and oldest method used in liposome technology. In this method, lipids are firstly dissolved in a suitable organic solvent, and dried down to yield a thin film at the bottom of the flask. The obtained lipid film is hydrated using an appropriate aqueous medium to produce liposomal dispersion. The structural organization of the formed vesicles can be affected by the hydration conditions. A gentle hydration of the lipid film forms giant unilamellar vesicles (GULV), whereas a hash hydration gives rise to multilamellar vesicles (MLV) with poor size homogeneity, which requires an additional downsizing step. The most commonly used sizing methods include probe and bath sonication that afford production of small unilamellar vesicles (SUV). Despite its higher effectiveness, probe sonication is often blamed for potential contamination (with titanium from the titanium-based nozzle used for mechanical agitation), and production of local heat that can affect lipids and drugs stability. Although the two sonication methods produce liposomes with identical characteristics, the use of bath sonication remains a better option due to easy control of operational parameters. Another technique used for liposome sizing includes consecutive extrusion of the liposomal formulation through polycarbonate filters of defined pore sizes. In this method, the number of extrusion cycles is the key parameter to control for effective homogenization [4, 6].

4.1.1.2 Reverse phase evaporation

Reverse phase evaporation is an alternative method to the film hydration that involves formation of water-in-oil emulsion between the aqueous phase (containing hydrophilic materials) and the organic phase (containing lipids and any hydrophobic materials). A brief sonication of this mixture is required for system homogenization. The removal of the organic phase under reduced pressure yields a milky gel that turns subsequently into liposomal suspension. The liposomes can be isolated from the dispersion using centrifugation, dialysis or sepharose 24 column [28].

4.1.1.3 Solvent injection

Solvent injection involves quick injection of the lipid solution (in ethanol or diethyl ether) into an aqueous medium. The experiment is performed either at room or at higher temperature (e.g., 60°C), depending on whether the organic solvent is water-miscible or not. The liposomes prepared by solvent injection process are mostly polydispersed and highly contaminated by organic solvents, especially ethanol due to formation of azeotrope mixture with water. As presented in **Table 3**, solvent injection suffers from several drawbacks including continuous exposure of the therapeutic agents to high temperature and organic solvent that might affect both the stability and safety of the liposomal products [28, 40].

4.1.1.4 Detergent removal

In the detergent removal method, phospholipids are dissolved in aqueous solution containing detergents at critical micelle concentrations (CMC). Upon detergent removal, the reaction medium frees individual phospholipid molecules that self-assemble into bilayered structures. Detergent removal is mostly achieved by means of a dialysis bag, polystyrene-based absorber beads or Sephadex columns (gel permeation chromatography). Dilution of the resultant mixture with some appropriate aqueous medium leads to restructuration of the formed micelles that evolve to liposomes [4, 6].

4.1.1.5 Heating method

Among all the conventional methods, the heating method is known to be the most attractive method for liposomes preparation due to its organic solvent free characteristics. In the heating method, lipids are hydrated for 1 hour, and heated for another hour above the transition temperature of the phospholipids in the presence of a hydrating agent (glycerin or propylene glycol 3%). When cholesterol is part of the formulation, the reaction medium is heated up to 100°C because of its high melting point. Being prepared under heating conditions, the resultant liposomes can be readily used without any further sterilization treatments, which minimizes both formulation complexity and timing. In addition, there is no need for further removal of the hydrating agents employed, since these represent physiologically acceptable ingredients that are well-established for pharmaceutical applications. Moreover, the observation that these hydrating agents can prevent particle coagulation and sedimentation makes them much more attractive as stabilizer and isotonizing additives. The hydroxyl groups of these hydrating agents provide a cryoprotective effect that makes the heating method an efficient method for the formulation of inhalable liposomes [41, 42].

4.1.2 Novel preparation methods

4.1.2.1 Microfluidic channel method

Microfluidic methods include all the novel techniques that make use of microscopic channels (in the size range of 5–500 μm). In this method, lipids are dissolved in an appropriate organic solvent (ethanol or isopropanol) and the resultant solution is propelled perpendicularly or in the opposite direction to the aqueous medium within the micro-channels. The continuous axial mixing of the organic and aqueous solutions leads to liposomes formation because of local diffusion of phospholipids in aqueous phase, which encourages the self-assembly process. Among many others, the micro hydrodynamic focusing method represents the most commonly used microfluidic method for liposomes formulation. This method produces small and large unilamellar vesicles, 40–140 nm, with good size homogeneity (mono dispersed feature). The other microfluidic techniques include the microfluidic droplets and the pulsed jet flow microfluidic methods. The microfluidic droplets method involves dissolution of phospholipids in hexane for preparation of giant liposomes (4–20 μm). In the pulsed jet flow microfluidic method, the conventional film hydration method has been modified by drying the lipid solution in microtubes. The resultant lipid film is hydrated within the microtubes through a perfusion process that produces much larger vesicles, 200–534 μm, with remarkable encapsulation efficiency [4, 43]. As common advantages, the microfluidic methods offer the possibility for production of vesicles with desired size, due to the versatility and flexibility of the methods. The disadvantages of these methods include the imperative use of organic solvent and drastic agitation, as well as difficulty for large scale production [40].

4.1.2.2 Supercritical fluidic method

While being considered as equivalent to the conventional reverse phase evaporation method, supercritical fluidic technique represents the most important novel liposome preparation methods that makes use of a supercritical fluid, such as carbon dioxide (CO_2) maintained under supercritical conditions (supercritical temperature and pressure). In this state, CO_2 is an excellent solvent for the lipids. The high-performance liquid chromatography (HPLC) pump provides a continuous flow of the aqueous phase into a view cell that contains the supercritical lipid solution, allowing phase transition of the dissolved phospholipids. Upon sudden decrease in pressure, CO_2 gets completely removed and phospholipids self-assemble into a bilayered vesicular system. The supercritical fluidic method affords large unilamellar vesicles (100–1200 nm) with 5-fold higher encapsulation efficiency than the equivalent conventional method. Apart from being organic solvent-free methods, the supercritical fluidic method offers many other advantages such as the use of CO_2, as a cheap and environmentally harmless solvent, possibility for controlling particle size, in situ sterilization and large-scale production in industrial settings. However, the disadvantages of the supercritical fluidic technique, including particularly its high cost, low yield and use of high pressures (200–350 bar) which require special infrastructures, restrict their universal applications for wider developments of liposomal technology [4, 40, 43].

4.1.3 Post-preparation treatments

4.1.3.1 Freeze-thawing

The freeze-thawing treatment involves freezing the liposomes dispersion in liquid nitrogen, and subsequently thawing it at the temperature above the phase transition temperature of the lipids used for formulation. Upon freeze-thawing, the liposomal vesicles are subjected to fusion since the lipid bilayers become fluid and highly permeable, allowing extensive diffusion of hydrophilic molecules, which leads to important cryoconcentration. These structural modifications encourage encapsulation of hydrophilic materials that are poorly loaded in liposomes when conventional methods are used. This underlines the reason why freeze-thawing represents an important treatment in liposome technology. Amongst the key parameters to be considered for freeze-thawing optimization are the number and duration of freeze-thawing cycles. These can impact significantly not only the encapsulation efficiency but also structural characteristics, i.e., liposomes lamellarity and polydispersity [44, 45].

4.1.3.2 Freeze-drying

Commonly known as lyophilization, freeze-drying is a post-preparation treatment for liposomes that is applied in both laboratory and industrial settings to preserve the characteristics of liposomal products. Freeze-drying involves freezing of the aqueous samples and subsequent removal of ice by sublimation. Freeze-drying represents a very useful treatment for shelf stability of liposomal suspensions, since water molecules can trigger some chemical reactions and lead to modification of the cargo or excipients in the formulation. Freeze-drying appears to be of great interest when the prepared formulation contains thermos-sensitive materials such as proteins, nucleic acids, etc., which might undergo fast degradation when subjected to heat-drying. The use of freeze-drying has gained considerable attention in liposome technology due to improved storage stability of liposomal products. Because of potential leakage of liposomes during freeze-drying, addition of hydrophilic compounds, commonly called cryoprotective agents (such as carbohydrates), has been established to ensure good stability and quality of the final product. The cryoprotectants commonly used include mannitol, lactose, sucrose and trehalose. Among these, trehalose is the most reputed cryoprotecting agent since it preserves liposomes stability during and after freeze-drying treatment [6].

4.2 Characterization techniques

After production, liposome formulations are subjected to extensive characterization, evaluating the physicochemical properties of liposomes that affect their shelf stability and biological performance. The most routinely investigated parameters in liposome characterization include vesicle size and size distribution (or polydispersity), surface charge (or Zeta potential), shape and morphology, lamellarity, encapsulation efficiency, phase behavior (or polymorphism) and in vitro release profile [4, 46]. **Table 4** indicates the techniques used for evaluation of liposome characteristics. The most frequently used methods are briefly discussed in the following paragraphs.

4.2.1 Dynamic light scattering (DLS)

Also known as photon correlation spectroscopy or quasi-elastic light scattering, DLS represents the most commonly used method for determination of liposome size, size distribution (polydispersity) and Zeta potential (surface charge). DLS is done by an instrumental setting called Zetasizer Nano. The standard operational principle of DLS is based on continuous motion of dispersed particles due to their bombardment by solvent molecules (Brownian motion). This phenomenon causes remarkable scattering of the applied light. Since the extent of fluctuation in light intensity is associated with the diffusion rate of the suspended particles, which is related to particle diameter (smaller particles diffusing faster than the larger ones), the particle size is automatically deducted from the estimated amount of the scattered light. When addressing Zeta potential measurements, DLS allows surface charge determination by accessing changes in the scattered light intensity caused by particle motion due to the electric field applied. In other words, for surface

Parameters	Analytical techniques
Particle size	Dynamic light scattering (DLS), size exclusion chromatography (SEC), field-flow fractionation (FFF) and microscope technology: transmission electron microscopy (TEM), cryogenic-TEM (Cryo-TEM), and atomic force microscopy (AFM)
Zeta potential /Surface charge	Electrophoretic mobility, DLS
Particle shape / morphology	Microscopic techniques such as TEM, Cryo-TEM and AFM
Lamellarity	Cryo-TEM and ^{31}P-NMR
Phase behavior	X-ray diffraction (XRD), differential scanning calorimetry (DSC) and thermogravimetric analysis (TGA)
Encapsulation efficiency	Centrifugation, dialysis or column separation for liposomes isolation, followed by drug content determination
Drug release	Dialysis or centrifugation, followed by drug quantification using analytical method, such as UV–Vis spectrophotometry, fluorescence spectrometry, enzyme- or protein-based assays, gel electrophoresis, HPLC, UPLC, LC-MS

Table 4.
Analytical methods commonly used for liposomes characterization.

charge (Zeta potential) evaluation, changes in the intensity of the scattered light are governed by the applied electric field (which causes extensive motion of charged particles), in contrast to size measurements where Brownian motion is the key factor [4, 47]. Apart from being a simple, fast and reliable method for routine analyses, DLS offers many other advantages including the fact that the measurement is taken from a native environment, and a wide size range can be evaluated (from a few nanometers to several micrometers). However, DLS shows some limitations such as the difficulty of differentiating individual particles from aggregates and high sensitivity to contaminants [48]. In addition, DLS is technically unable to provide true particle size, but rather hydrodynamic diameter due to particle solvation. Water layers on particle surface lead to false readings of particle diameters in aqueous media [49].

4.2.2 Transmission electron microscopy (TEM)

The microscopic observation provides direct visualization of the liposomal vesicles as individual particles, which allows effective analysis of shape and morphology as well as a precise and reliable size reading. In this context, TEM techniques are commonly used in liposome technology for structural characterizations. In TEM experiments, the liquid sample is spotted onto a copper grid, and the solvent dried prior to the microscopic analysis. Under TEM instrument, liposomal vesicles mostly appear as black spherical particles on a white background. For a variant TEM technique like negative staining TEM, liposomes appear as bright spherical spots on a black background since the spotted sample is treated with uranyl acetate or phosphotungstic acid (as negative staining agent). Due to its effectiveness, TEM appears to be a powerful complementary technique to DLS for confirmation of the liposomal structure. Unlike DLS, TEM offers the advantage of differentiating individual vesicles from aggregates, allowing critical assessment of the liposome population. Nevertheless, TEM presents several limitations due to sample preparation. Apart from being time-consuming, sample pretreatments in TEM analyses may cause remarkable changes in liposomal shape or morphology: potential vesicle shrinkage, swelling or artifact formation [4, 50].

To overcome these limitations, Cryo-TEM was developed as a strategy to minimize liposome disruption by making use of a flash freezing treatment for direct particle visualization in solid-state (without solvent removal). Nowadays, Cryo-TEM is the most reliable technique for visual determination of liposome structure including lamellarity. However, Cryo-TEM appears to be sometimes limited since it works perfectly only with very small particles. This has led to the development of atomic force microscopy (AFM) for direct particle analysis in native environments. Although AFM offers the advantage of higher particle resolution at three-dimensional level, the use of this technique is mostly limited by the high cost of the instrument, which compromises its universal availability and accessibility [3, 4].

4.2.3 Differential scanning calorimetry (DSC) and X-ray diffraction (XRD)

DSC and XRD are complementary techniques that evaluate the thermal behavior and crystallinity, respectively, and provide valuable information for characterization of loaded liposomes [46, 51, 52]. DSC evaluates the differences in heat flow (electric power) between a sample and a reference. In DSC experiments, the sample and the reference are subjected to a programmed heating, cooling or isothermal treatment in a controlled atmosphere (mostly saturated with nitrogen gas). The heating treatment is achieved either by the same heater (heat flux DSC) or by separate heaters (power compensated DSC). The experiment is conducted in specialized metal pans made of aluminum, tin, zinc or indium. Throughout the experiment, frequent electric power adjustments occur upon material phase transition (melting or crystallization), ensuring thermal equilibrium between the sample and the reference. This phenomenon is described and expressed by the plot of heat capacity against temperature or time (heat flow curve). The heat flow curve provides the respective transition temperature and enthalpy, which allows to identify the nature of thermal events: endo- or exothermic [53–55]. DSC represents the most useful thermal analysis technique in the study of lipid-based materials [46, 52, 54–56].

Unlike DSC where sample recovery is not possible, XRD is a non-destructive analytical tool that allows structural investigations of crystalline materials. XRD makes use of X-rays that deeply penetrate solid materials and provide useful information at atomic structure level. Though relatively expensive, an XRD instrument is an environmentally and user-friendly device that is easy to use. A wide range of materials such as powders, crystals and liquids can be quickly assessed by XRD. Its other advantages include high resolution, reliability, relatively cheap maintenance, and easy data collection, processing and interpretation. The phase transitions and polymorphism determined by XRD represent valuable information in pharmaceutical development and production of both excipients and biologically active materials [57–59].

4.2.4 Lamellarity assays

The lamellarity of liposomes is part of their structural characteristics that can have an impact on the intended applications. The number of lipid bilayers can be evaluated using chemically labeled or radiolabelled agents that can be distributed in the bilayer membranes. However, this technique is limited since these reagents might be distributed only on the outer lipid membrane and lead to false readings. To date, Cryo-TEM is the most commonly used technique for the determination of lamellarity by visualization [4]. The nuclear magnetic resonance spectroscopy of the 31-phosphorus (^{31}P-NMR) is also being currently used to estimate the lamellarity of liposomes. This technique deals with the estimation of the ratio of phospholipid amount in the outer layers to that of the inner layers [60]. The ^{31}P-NMR spectrum with a broad peak indicates the presence of MLV while a narrow peak corresponds to SUV. The addition of paramagnetic ions such as Mn^{2+}, Co^{2+}, and Pr^{3+} shifts the respective peaks to either downfield or upfield because of ionic interactions with the phosphate backbone. By comparing the spectroscopic profile with and without the paramagnetic ion, the lamellarity of liposomes can be estimated. Some other techniques such as small-angle X-ray scattering (SAXS) and trapped volume determination can be used to estimate liposome lamellarity [4, 61].

4.2.5 In vitro release assays

The profile of release for the liposomal cargoes is commonly estimated in vitro using dialysis. This method implies trapping the liposomal dispersion into a dialysis bag of specific molecular weight cut off. The resultant tubing membrane is placed in a simulated physiological fluid (release medium) that is often a buffer maintained under well-defined conditions: specific temperature and speed of stirring/shaking. At predetermined time intervals, an aliquot is withdrawn from the release medium and an equal volume of the fresh buffer is replaced to maintain sink conditions. In the withdrawn sample aliquots, the released cargo is quantified using some routine analytical techniques such as UV-Vis spectrophotometry, HPLC, UPLC, etc., adapted to the molecular species under evaluation. The release profile is obtained by plotting the cumulative release percentage against the chosen time intervals [4]. Data from the in vitro release study are valuably considered as part of the rational development of formulations for controlled release, since they allow effective prediction of in vivo performance of the delivery systems [62].

5. Applications

Liposomes have evolved so far from mere experimental tools of research to industrially established products for clinical and veterinary use. They have shown the ability to improve the physicochemical features of the cargoes and ferry them to the sites of interest. The concepts of liposomal encapsulation have been applied in several fields of life science. Liposomes are frequently used for the delivery of drug, gene, vaccine and diagnostic products; but other applications encompass encapsulation of food and cosmetic ingredients as well as routine analysis of chemical substances [2, 3]. The following paragraphs briefly present the current applications of liposomes.

5.1 Application in drug delivery

The use of liposomes in drug delivery aims at modifying the pharmacokinetics of drugs to improve the therapeutic efficacy while minimizing potential toxicity [6]. Liposomes can alter the spatial and temporal distribution of the entrapped drug molecules in vivo, leading to controlled delivery at the site of interest and reduced off-target adverse effects [63]. The liposomal systems have been extensively investigated for the delivery of existing and emerging drugs at various research levels, from basic stages related to research and development to preclinical and clinical applications. Nowadays, liposomes represent the most clinically established drug vehicles for human diseases [3, 5]. The efforts invested in liposomal technology have so far led to the development of several effective liposomal formulations that are currently used in clinics (**Table 5**).

Liposomal formulations have been used to address a wide range of pathological conditions through different administration routes, including dermal, transdermal, oral, pulmonary and parenteral routes. The clinical areas commonly explored in liposome research encompass skin disorders, cancers and infectious diseases [4].

Branded product	Drug name	Therapeutic indications
Abelcet	Amphotericin B	Fungal infections
AmBisome	Amphotericin B	Fungal infections
Amphocil	Amphotericin B	Fungal infections
DaunoXome	Daunorubicin	Hematological malignancy
DepoCyt	Cytarabine	Lymphomatous meningitis
DepoDur	Morphine sulfate	Pain relief
Doxil	Doxorubicin	Kaposi's sarcoma and solid tumors
Epaxal	Inactivated hepatitis A virus	Hepatitis A
Evacet	Doxorubicin	Ovarian cancer
Inflexal V	Inactivated hemagglutinin of influenza virus strains A and B	Influenza
LipoDox	Doxorubicin	Kaposi's sarcoma and solid tumors
Marqibo	Vincristine sulfate	Acute lymphoblastic leukemia
Visudyne	Verteporfin	Photodynamic therapy

Table 5.
Clinically approved liposomal products [3, 4].

Amongst, cancer therapy appears to be in the forefront of liposome delivery, due to poor bioavailability and side effects of most of the anti-cancer drugs. However, several infectious diseases, most specially where the pathogen is hosted by the MNPS (i.e., tuberculosis, leishmaniasis, fungal infections), have been reported to be good candidates for liposome application, taking advantage of the spontaneous liposomes uptake by the cells of MNPS [2]. Apart from the nature and localization of the disease, the design and development of liposomes depend also on the intended administration route, since different anatomical and physiological characteristics can be encountered from one route to another [46].

Furthermore, the application of liposomes in drug delivery is highly dependent on their colloidal and physiochemical features, i.e., vesicle size, surface charge and system stability [64]. For instance, small liposomes (SUV) are good candidates for Parkinson's and Alzheimer's diseases, due to the need for crossing the brain blood barrier to achieve brain targeted delivery. Meanwhile, large liposomes are preferred for macrophage targeted delivery of antimicrobials, when pathogens are located inside the MNPS cells (e.g., tuberculosis, leishmaniasis), taking advantage of the passive liposome cell uptake [2, 4, 46]. This underlines the need for thorough exploration of process and formulation parameters at early stages of products development to produce liposomes with desired characteristics, making the technology for liposomes manufacturing key to future therapeutic research and development.

5.2 Application in gene delivery

Liposomes have been reported to achieve effective intracellular delivery of genes. These liposomes, also called lipoplexes, are generally made of cationic lipids, which allow for encapsulation of genetic materials via electrostatic interactions with the negatively charged phosphate backbones of nucleic acids. The positive charge on the surface of liposomes also influences their interactions with negatively charged cell membranes and promotes cell internalization [23, 65]. Cationic liposomes offer several advantages over viral gene vectors, including easy and safe production, cost effectiveness, possibility of monitoring toxicity, biodegradability, biocompatibility and lack of dangerous immunogenicity. However, they are suffering from poor transfection efficiency due to their limited endosomal escaping ability, which exposes the genetic materials to enzymatic and acid degradation in lysosomes. This has been improved by incorporating fusogenic lipids, such as DOPE, in the liposomal formulation. These helper lipids facilitate endosomal escaping by membrane fusion with endosomes, leading to early cytoplasmic release of the gene. Though cationic liposomes have shown some dose-dependent toxicity, successful results have been obtained in cancer therapy when delivering genes encoding for tumor suppression proteins [66, 67]. He et al. have recently developed folate receptor alpha-targeted lipoplexes with therapeutic gene expression regulated by an hTERT promoter. These liposomes have shown some promise for the treatment of ovarian cancer [68].

5.3 Application in vaccine delivery

The use of specific lipids or molecules such as phosphatidylserine, DOTAP, fatty acids and monophosphoryl lipids can produce liposomes with attractive immune-stimulating activities. Liposomes carrying antigenic materials, either encapsulated in the aqueous core, grafted or coated on the surface; can stimulate immune responses on macrophage uptake. Following endolysosomal degradation, macrophages present the antigen to T-lymphocytes that initiate the production of cytokines and specific antibodies via activation of B-lymphocytes [4]. The immune response produced by liposomes can be influenced by their composition, lamellarity, size and surface charge [69]. Liposomes containing a glycolipid, trehalose 6,6'-dibehenate, and a cationic lipid, dimethyldioctadecylammonium, in a 1:5 mass ratio have demonstrated efficient delivery of the TB vaccine Ag85B-ESAT-6. This vaccine has shown prolonged immune response without any toxic effects [70]. Although a liposomal vaccine (namely Stimuvax®) targeting the major histocompatibility class I complex for lung carcinoma was not successfully implemented, some other vaccines such as Epaxal®, Inflexal® V and Mosquirix® have been clinically established for the treatment of hepatitis A, influenza virus infections and malaria, respectively. These vaccines, classified as virosomes, are liposomes generally made of reconstituted viral envelop supplemented with phosphatidylcholine. They offer the advantage of undergoing membrane fusion either with the cells or the endosomes, and thus leading to efficient cytosolic delivery [2, 9].

5.4 Application in diagnosis

The use of liposomes for diagnostic purposes is one of current topics of great interest in biomedical applications. Liposomes with magnetic properties, also called magnetoliposomes, are made by entrapping superparamagnetic iron-based nanoparticles or iron oxides or gadolinium (III) chelates for magnetic resonance imaging. Functionalized liposomes labeled with radioisotopes can also be used as molecular probes in nuclear imaging. For instance, liposomes labeled with ^{64}Cu, ^{18}F, ^{89}Zr or ^{52}Mn have been reported in positron emission tomography while ^{99m}TC, ^{111}In or ^{67}Ga labeled liposomes were applied in single photon emission computed tomography. Acoustic liposomes which are liposomes made of perfluoropropane gas can be used as contrast agents in ultrasound imaging technique. The encapsulation of quantum dots and fluorescent dyes into liposomes has also led to the development of attractive liposomal platforms for diagnosis. Additionally, liposomes have shown great potential for simultaneous accommodation of drugs and diagnostic agents such as radionuclides, magnetic or contrast substances. This can be achieved by encapsulation in the inner core, embedding in the lipid bilayer, chemically grafting or coating onto the surface of liposomes. A successful targeted co-delivery of these materials has given to liposomes the status of theranostic systems, as they provide both the therapeutic effect and the diagnostic control [2, 71, 72].

5.5 Application in analytical fields

In analytical domains, liposome-based formulations can be usefully involved in immunoassays, biosensors analysis and liquid chromatography. Liposomes can be used to encapsulate, embed or conjugate the analytical entities with high loading capacity and huge surface area that can enhance the intensity of analytical signal. In a direct enzyme-linked immunosorbent assay (ELISA) like method, fluorophore-conjugated liposomes carrying a substantial amount of secondary antibody quickly bind to the antigen that has been fixed on the primary antibody. The addition of a colorimetric substrate leads to color development and allows for analytical estimation of the antigenic analyte. The use of liposomes has been deemed to lower the limits of detection of analytes and increase the sensitivity of immunological analysis technique [73]. In chromatography, the conjugation of liposomes to the stationary phase for gel permeation chromatography is useful for separation of drugs and proteins and for exploring possible molecular interactions on phospholipid membranes. Additionally, the cell-like appearance of liposomes makes them appropriate simulated cell models for studying and predicting the interactions between biologically active compounds and cell membranes [4].

5.6 Application in cosmetics

The use of liposomes in the field of cosmetics is based on the similarity between the lipid composition of the liposomal vesicle and that of the biological layers composing the skin. Because phospholipids can be subject to hydration, topical liposomes happen to contribute to the reduction of skin dryness. While playing the role of attractive vehicles for relevant cosmetic agents, liposomes provide a great source of skin ingredients, such as essential unsaturated fatty acids like linolenic acid. This has shown great potential for maintaining the skin and hair in good physiological standing, preventing the rise of some common topical diseases. Furthermore, the use of skin care formulations made of empty or hydrating agents loaded liposomes helps to reduce the transdermal water loss, which is a major cause of skin dryness [74, 75].

5.7 Application in food industry

As versatile lipid-based systems, liposomes have shown some potential in the encapsulation of food ingredients. In this field, the use of liposomes aims to stabilize some nutraceutical or dietetic ingredients during the storage, to improve their organoleptic characters or to provide a controlled and targeted delivery of these substances in a specific tissue. The shelf life and efficacy of instable bioactive products such as vitamins, enzymes and anti-oxidative agents have been prolonged by their liposomal encapsulation. For example, proteinase and lipase loaded liposomes have been reported to improve the ripening of cheese notably in preventing proteolysis of casein [28, 76]. Yokota et al. [27] have successfully improved the taste and odor of casein hydrolysate in dietetic preparations by liposomal encapsulation. In addition to the encapsulation of dietetic compounds, liposomes have been recently used to encapsulate a cyanobacterium, namely Spirulina platensis, as a source of proteins [77]. Although the use of liposomes in food industry is still at the infant stages of development, advanced studies have been launched to investigate much more attractive applications, such as detection of food contaminants by means of stimuli-responsive liposomes for food safety [76].

6. Conclusion

Liposomes appear to be reputed carriers for various chemical and macromolecular species. From way back their discovery, liposomes have been subject to extensive evolution, in terms of composition, manufacturing and usages, which led to several openings in both basic and applied life sciences. The general details presented herein attempted to reveal some of the existing gaps in liposomes technology and open new windows for further research. An important future breakthrough could be the discovery of cost-effective materials to formulate liposome vesicles with remarkable versatility viz., being suitable as vehicles for various molecules while holding a clear molecular architecture to act as cell models for in vitro bioassays. In addition, novel manufacturing methods for facile encapsulation of both hydrophobic and hydrophilic molecules with no need for organic solvents and special equipment or sophisticated infrastructures are highly desired. Furthermore, the critical issue of system instability for liposomes loaded with hydrophilic materials requires new strategies that will achieve acceptable loading while aiming at targeting cargo delivery at the site of interest. Finally, since liposomes have demonstrated some clinical success as drug vehicles, future efforts should be dedicated to ensuring wider developments and implementation of therapeutic liposomes. This would enhance commercial availability and accessibility of liposome products across the globe, particularly in low- and middle-income countries.

Acknowledgements

We acknowledge the support received from the National Research Foundation of South Africa and the *Sandisa Imbewu* Programme of Rhodes University. CI Nkanga and AM Bapolisi are grateful to the NGO *Förderverein Uni Kinshasa e. V.-BEBUC/ Else-KroenerFresenius Stiftung* and *Holger-Poehlmann* for the advice.

Conflict of interest

There is no conflict of interest to declare.

Author details

Christian Isalomboto Nkanga, Alain Murhimalika Bapolisi,
Nnamdi Ikemefuna Okafor and Rui Werner Maçedo Krause*
Department of Chemistry, Centre for Chemico- and Bio-Medicinal Research (CCBR), Rhodes University, South Africa

*Address all correspondence to: r.krause@ru.ac.za

References

[1] Bangham AD, Standish MM, Watkins JC. Diffusion of univalent ions across the lamellae of swollen phospholipids. Journal of Molecular Biology. 1965;**13**:IN26-IN27. DOI: 10.1016/S0022-2836(65)80093-6

[2] Abu Lila AS, Ishida T. Liposomal delivery systems: Design optimization and current applications. Biological & Pharmaceutical Bulletin. 2017;**40**:1-10. DOI: 10.1248/bpb.b16-00624

[3] Bozzuto G, Molinari A. Liposomes as nanomedical devices. International Journal of Nanomedicine. 2015;**10**: 975-999. DOI: 10.2147/IJN.S68861

[4] Pattni BS, Chupin VV, Torchilin VP. New developments in liposomal drug delivery. Chemical Reviews. 2015;**115**:10938-10966. DOI: 10.1021/acs.chemrev.5b00046

[5] Bulbake U, Doppalapudi S, Kommineni N, Khan W. Liposomal formulations in clinical use: An updated review. Pharmaceutics. 2017;**9**:1-33. DOI: 10.3390/pharmaceutics9020012

[6] Akbarzadeh A, Rezaei-Sadabady R, Davaran S, Joo SW, Zarghami N, Hanifehpour Y, et al. Liposome: Classification, preparation and applications. Nanoscale Research Letters. 2013;**8**:1-8. DOI: 10.1186/1556-276X-8-102

[7] Randles EG, Bergethon PR. A photodependent switch of liposome stability and permeability. Langmuir. 2013;**29**:1490-1497. DOI: 10.1021/la303526k

[8] Tapaswi RD. Liposome as a potential drug delivery system: A review. International Research Journal of Pharmacy. 2013;**4**:6-12. DOI: 10.1152/physiol.00045.2008

[9] Kamboj S, Saini V, Magon N, Bala S, Jhawat V. Vesicular drug delivery systems: A novel approach for drug targeting. International Journal of Drug Delivery. 2013;**5**:121-130

[10] Ghanbarzadeh S, Arami S. Enhanced transdermal delivery of diclofenac sodium via conventional liposomes, ethosomes, and transfersomes. BioMed Research International. 2013;**7**. DOI: 10.1155/2013/616810

[11] Yu X, Du L, Li Y, Fu G, Jin Y. Improved anti-melanoma effect of a transdermal mitoxantrone ethosome gel. Biomedicine & Pharmacotherapy. 2015;**73**:6-11. DOI: 10.1016/j.biopha.2015.05.002

[12] Bomsel M, Tudor D, Drillet A, Alfsen A, Ganor Y, Roger M, et al. Immunization with HIV-1 gp41 subunit virosomes induces mucosal antibodies protecting nonhuman primates against vaginal shiv challenges. Immunity. 2011;**34**:269-280

[13] Kaur K, Kush P, Pandey RS, Madan J, Jain UK, Katare OP. Stealth lipid coated aquasomes bearing recombinant human interferon-α-2b offered prolonged release and enhanced cytotoxicity in ovarian cancer cells. Biomedicine & Pharmacotherapy. 2015;**69**:267-276. DOI: 10.1016/j.biopha.2014.12.007

[14] Shukla A, Singh B, Katare OP. Significant systemic and mucosal immune response induced on oral delivery of diphtheria toxoid using nano-bilosomes. British Journal of Pharmacology. 2011;**164**:820-827. DOI: 10.1111/j.1476-5381.2011.01452.x

[15] Yuksel N, Bayindir ZS, Aksakal E, Ozcelikay AT. In situ niosome forming maltodextrin proniosomes of candesartan cilexetil: In vitro and in vivo evaluations. International Journal of Biological Macromolecules. 2016;**82**:453-463. DOI: 10.1016/j.ijbiomac.2015.10.019

[16] Li J, Wang X, Zhang T, Wang C, Huang Z, Luo X, et al. A review on phospholipids and their main applications in drug delivery systems. Asian Journal of Pharmaceutical Sciences. 2014;**10**:81-98. DOI: 10.1016/j.ajps.2014.09.004

[17] Patel N, Panda S. Liposome drug delivery system: A critic review. Journal of Pharmaceutical and BioSciences Resaearch. 2012;**2**:162-194

[18] Vemuri S, Rhodes CT. Preparation and characterization of liposomes as therapeutic delivery systems: A review. Pharmaceutica Acta Helvetiae. 1995;**70**:95-111

[19] Taylor G, Taylor G, Kellaway I, Stevens J. Drug entrapment and release from multilamellar and reverse-phase evaporation liposomes. International Journal of Pharmaceutics. 1990;**58**:49-55

[20] de Kruyff B, Demel RA, van Deenen LL. The effect of cholesterol and epicholesterol incorporation on the permeability and on the phase transition of intact Acholeplasma laidlawii cell membranes and derived liposomes. Biochimica et Biophysica Acta. 1971;**255**:331-347

[21] Kirby C, Clarke J, Gregoriadis G. Effect of the cholesterol content of small unilamellar liposomes on their stability *in vivo* and *in vitro*. The Biochemical Journal. 1980;**186**:591-598. DOI: 10.1042/bj1860591

[22] López-Pinto JM, González-Rodríguez ML, Rabasco AM. Effect of cholesterol and ethanol on dermal delivery from DPPC liposomes. International Journal of Pharmaceutics. 2005;**298**:1-12. DOI: 10.1016/j.ijpharm.2005.02.021

[23] Paecharoenchai O, Niyomtham N, Apirakaramwong A, Ngawhirunpat T, Rojanarata T, Yingyongnarongkul B, et al. Structure relationship of cationic lipids on gene transfection mediated by cationic liposomes. AAPS PharmSciTech. 2012;**13**:1302-1308. DOI: 10.1208/s12249-012-9857-5

[24] Xu H, Deng Y, Chen D, Qin J, Liu J, Shi L, et al. Preparation and characterization of pH-sensitive vesicles made of cholesteryl hemisuccinate. Drug Development and Industrial Pharmacy. 2008;**34**:134-141. DOI: 10.1080/03639040701410903

[25] Hafez IM, Cullis PR. Cholesteryl hemisuccinate exhibits pH sensitive polymorphic phase behavior. Biochimica et Biophysica Acta. 2000;**1463**:107-114. DOI: 10.1016/S0005-2736(99)00186-8

[26] Xu H, Deng Y-H, Wang K-Q, Chen D-W. Preparation and characterization of stable pH-sensitive vesicles composed of α-Tocopherol Hemisuccinate. AAPS PharmSciTech. 2012;**13**:1377-1385. DOI: 10.1208/s12249-012-9863-7

[27] Yokota D, Moraes M, Pinho SC. Characterization of lyophilized liposomes produced with non-purified soy lecithin: A case study of casein hydrolysate microencapsulation. Brazilian Journal of Chemical Engineering. 2012;**29**:325-335. DOI: 10.1590/S0104-66322012000200013

[28] Machado AR, Leticia M d A, Maria IRM, Leonor A d S-S. Importance of lecithin for encapsulation processes. African Journal of Food Science. 2014;**8**:176-183. DOI: 10.5897/AJFS2013.1092

[29] Nkanga CI, Krause RW, Noundou XS, Walker RB. Preparation and characterization of isoniazid-loaded crude soybean lecithin liposomes. International Journal of Pharmaceutics. 2017;**526**:466-473. DOI: 10.1016/j.ijpharm.2017.04.074

[30] Immordino ML, Dosio F, LC. Stealth liposomes: Review of the basic science, rationale, and clinical applications,

existing and potential. International Journal of Nanomedicine. 2006;**1**: 297-315. DOI: 10.1023/A:1020134521778

[31] Torchilin VP. Recent advances with liposomes as pharmaceutical carriers. Nature Reviews. Drug Discovery. 2005;**4**:145-160. DOI: 10.1038/nrd1632

[32] Nisini R, Poerio N, Mariotti S, De Santis F, Fraziano M. The multirole of liposomes in therapy and prevention of infectious diseases. Frontiers in Immunology. 2018;**9**. DOI: 10.3389/fimmu.2018.00155

[33] Drummond DC, Zignani M, Leroux JC. Current status of pH-sensitive liposomes in drug delivery. Progress in Lipid Research. 2000;**39**:409-460. DOI: 10.1016/S0163-7827(00)00011-4

[34] Karanth H, Murthy RSR. pH-sensitive liposomes-principle and application in cancer therapy. The Journal of Pharmacy and Pharmacology. 2007;**59**:469-483. DOI: 10.1211/jpp.59.4.0001

[35] Cheng FF, Zhang JJ, Xu F, Hu LH, Abdel-Halim ES, Zhu JJ. PH-sensitive polydopamine nanocapsules for cell imaging and drug delivery based on folate receptor targeting. Journal of Biomedical Nanotechnology. 2013;**9**:1155-1163. DOI: 10.1166/jbn.2013.1611

[36] Bhardwaj A, Grobler A, Goutam R, Goyal AK, Jain AK, Mehta A. Pulmonary delivery of anti-tubercular drugs using ligand anchored pH-sensitive liposomes for the treatment of pulmonary tuberculosis. Current Drug Delivery. 2016;**13**:909-922

[37] Nkanga CI, Walker RB, Krause RW. pH-dependent release of isoniazid from isonicotinic acid (4-hydroxy-benzylidene)-hydrazide loaded liposomes. Journal of Drug Delivery Science and Technology. 2018;**45**: 264-271. DOI: 10.1016/j.jddst.2018.03.016

[38] Nkanga CI, Krause RWM. Conjugation of isoniazid to a zinc phthalocyanine via hydrazone linkage for pH-dependent liposomal controlled release. Applied Nanoscience. 2018;**8**:1313-1323. DOI: 10.1007/s13204-018-0776-y

[39] Gharib R, Greige-Gerges H, Fourmentin S, Charcosset C, Auezova L. Liposomes incorporating cyclodextrin-drug inclusion complexes: Current state of knowledge. Carbohydrate Polymers. 2015;**129**: 175-186. DOI: 10.1016/j.carbpol.2015.04.048

[40] Maherani B, Arab-Tehrany E, R. Mozafari M, Gaiani C, Linder M. Liposomes: A review of manufacturing techniques and targeting strategies. Current Nanoscience. 2011;**7**:436-452. DOI: 10.2174/157341311795542453

[41] Mozafari R. Liposomes: An overview of manufacturing techniques. Cellular & Molecular Biology Letters. 2005;**10**:711-719

[42] Laouini A, Jaafar-Maalej C, Limayem-Blouza I, Sfar S, Charcosset C, Fessi H. Preparation, characterization and applications of liposomes: State of the art. Journal of Colloid Science and Biotechnology. 2012;**1**:147-168. DOI: 10.1166/jcsb.2012.1020

[43] Patil YP, Jadhav S. Novel methods for liposome preparation. Chemistry and Physics of Lipids. 2014;**177**:8-18. DOI: 10.1016/j.chemphyslip.2013.10.011

[44] Supaporn S. Effect of freeze-thawing process on the size and Lamellarity of PEG-lipid liposomes. The Open Colloid Science Journal. 2011;**4**: 1-8. DOI: 10.2174/1876530001104010001

[45] Costa AP, Xu X, Burgess DJ. Freeze-anneal-thaw cycling of unilamellar liposomes: Effect on encapsulation efficiency. Pharmaceutical Research.

2014;**31**:97-103. DOI: 10.1007/s11095-013-1135-z

[46] Prathyusha K, Muthukumaran M, Krishnamoorthy B. Reviw article liposomes As Targetted drug delivery systems present and future Prospectives : A review preparation passive loading mechanical dispersion methods solvent dispersion methods detergent. Journal of Drug Delivery & Therapeutics. 2013;**3**:195-201

[47] Hadinoto K, Sundaresan A, Cheow WS. Lipid-polymer hybrid nanoparticles as a new generation therapeutic delivery platform: A review. European Journal of Pharmaceutics and Biopharmaceutics. 2013;**85**:427-443. DOI: 10.1016/j.ejpb.2013.07.002

[48] Fissan H, Ristig S, Kaminski H, Asbach C, Epple M. Comparison of different characterization methods for nanoparticle dispersions before and after aerosolization. Analytical Methods. 2014;**6**:7324-7334. DOI: 10.1039/c4ay01203h

[49] Eaton P, Quaresma P, Soares C, Neves C, de Almeida MP, Pereira E, et al. A direct comparison of experimental methods to measure dimensions of synthetic nanoparticles. Ultramicroscopy. 2017;**182**:179-190. DOI: 10.1016/j.ultramic.2017.07.001

[50] Chetanachan P, Akarachalanon P, Worawirunwong D, Dararutana P, Bangtrakulnonth A, Bunjop M, et al. Ultrastructural characterization of liposomes using transmission electron microscope. Advances in Materials Research. 2008;**55-57**:709-711. DOI: 10.4028/www.scientific.net/AMR.55-57.709

[51] Singh C, Koduri LVSK, Singh A, Suresh S. Novel potential for optimization of antitubercular therapy: Pulmonary delivery of rifampicin lipospheres. Asian Journal of Pharmaceutical Sciences. 2015;**10**:549-562. DOI: 10.1016/j.ajps.2015.08.003

[52] Omwoyo WN, Ogutu B, Oloo F, Swai H, Kalombo L, Melariri P, et al. Preparation, characterization, and optimization of primaquine-loaded solid lipid nanoparticles. International Journal of Nanomedicine. 2014;**9**: 3865-3874. DOI: 10.2147/IJN.S62630

[53] Giron D. Applications of thermal analysis and coupled techniques in pharmaceutical industry. Journal of Thermal Analysis and Calorimetry. 2002;**68**:335-357. DOI: 10.1023/A:1016015113795

[54] Koyama TM, Stevens CR, Borda EJ, Grobe KJ, Cleary DA. Characterizing the gel to liquid crystal transition in lipid-bilayer model systems. The Chemical Educator. 1999;**4**:12-15. DOI: 10.1007/s00897990273a

[55] Ohline SM, Campbell ML, Turnbull MT, Kohler SJ. Differential Scanning Calorimetric Study of Bilayer Membrane Phase Transitions: A Biophysical Chemistry Experiment. Journal of Chemical Education. 2001;**78**(9):1251-2156

[56] Castelli F, Puglia C, Sarpietro MG, Rizza L, Bonina F. Characterization of indomethacin-loaded lipid nanoparticles by differential scanning calorimetry. International Journal of Pharmaceutics. 2005;**304**:231-238. DOI: 10.1016/j.ijpharm.2005.08.011

[57] Dorofeev GA, Streletskii AN, Povstugar IV, Protasov AV, Elsukov EP. Determination of nanoparticle sizes by X-ray diffraction. Colloid Journal. 2012;**74**:675-685. DOI: 10.1134/S1061933X12060051

[58] Gohil KN, Patel PM, Patel NM. Application of analytical techniques in preformulation study: A review. International Journal

of Pharmaceutical and Biological Archieves. 2011;**2**:1319-1326

[59] Chauhan A, Chauhan P. Powder XRD technique and its applications in science and technology. Journal of Analytical & Bioanalytical Techniques. 2014;**5**. DOI: 10.4172/2155-9872.1000212

[60] Frohlich M, Brecht V, Peschka-Suss R. Parameters influencing the determination of liposome lamellarity by ^{31}P-NMR. Chemistry and Physics of Lipids. 2001;**109**:103-112

[61] Bouwstra JA, Gooris GS, Bras W, Talsma H. Small angle X-ray scattering: Possibilities and limitations in characterization of vesicles. Chemistry and Physics of Lipids. 1993;**64**:83-98. DOI: 10.1016/0009-3084(93)90059-C

[62] Dash S, Murthy NP, Nath L, Chowdhury P. Kinetic modeling on drug release from controlled drug delivery systems. Acta Poloniae Pharmaceutica. 2010;**67**(2010):217-223. DOI: 10.2307/3237001

[63] Daraee H, Etemadi A, Kouhi M, Alimirzalu S, Akbarzadeh A. Application of liposomes in medicine and drug delivery. Artificial Cells, Nanomedicine, and Biotechnology. 2016;**44**:381-391. DOI: 10.3109/21691401.2014.953633

[64] Cagdas M, Sezer ADS, Bucak S. Liposomes as Potential Drug Carrier Systems for Drug Delivery. In: Application of Nanotechnology in Drug Delivery, Ali Demir Sezer, IntechOpen; 2014. DOI: 10.5772/58459. Available from: https://www.intechopen.com/books/application-of-nanotechnology-in-drug-delivery/liposomes-as-potential-drug-carrier-systems-for-drug-delivery

[65] Balazs DA, Godbey W. Liposomes for use in gene delivery. Journal of Drug Delivery. 2011;**2011**:1-12. DOI: 10.1155/2011/326497

[66] Haritha PN, Uma SKD, Nagaratna DP, Chaitanya SPK, Gunasekharan V. Gene therapy—A review. International Journal of Biopharmaceutics. 2012;**3**(1):55-64

[67] Dizaj SM, Jafari S, Khosroushahi AY. A sight on the current nanoparticle-based gene delivery vectors. Nanoscale Research Letters. 2014;**9**:1-9. DOI: 10.1186/1556-276X-9-252

[68] He ZY, Deng F, Wei XW, Ma CC, Luo M, Zhang P, et al. Ovarian cancer treatment with a tumor-targeting and gene expression-controllable lipoplex. Scientific Reports. 2016;**6**:1-13. DOI: 10.1038/srep23764

[69] Hussain MJ, Wilkinson A, Bramwell VW, Christensen D, Perrie Y. Th1 immune responses can be modulated by varying dimethyldioctadecylammonium and distearoyl-sn-glycero-3-phosphocholine content in liposomal adjuvants. The Journal of Pharmacy and Pharmacology. 2014;**66**:358-366. DOI: 10.1111/jphp.12173

[70] van Dissel JT, Joosten SA, Hoff ST, Soonawala D, Prins C, Hokey DA, et al. A novel liposomal adjuvant system, CAF01, promotes long-lived *Mycobacterium tuberculosis*-specific T-cell responses in human. Vaccine. 2014;**32**:7098-7107. DOI: 10.1016/j.vaccine.2014.10.036

[71] Xing H, Hwang K, Lu Y. Recent developments of liposomes as nanocarriers for theranostic applications. Theranostics. 2016;**6**: 1336-1352. DOI: 10.7150/thno.15464

[72] Lamichhane N, Udayakumar TS, D'Souza WD, Simone CB, Raghavan SR, Polf J, et al. Liposomes: Clinical applications and potential for image-guided drug delivery. Molecules. 2018;**23**:1-17. DOI: 10.3390/molecules23020288

[73] Singh AK, Kilpatrick PK, Carbonell RG. Application of antibody and fluorophorederivatized liposomes

to heterogeneous immunoassays for d-dimer. Biotechnology Progress. 1996;**12**(2):272-280

[74] Pradhan L, Kumar B, Saha N, Roy A. Liposome: Method of preparation, advantages, evaluation and its application. Journal of Applied Pharmaceutical Research. 2015;**3**(3):2348-0335

[75] Patravale VB, Mandawgade SD. Novel cosmetic delivery systems: An application update. International Journal of Cosmetic Science. 2008;**30**:19-33. DOI: 10.1111/j.1468-2494.2008.00416.x

[76] Shukla S, Haldorai Y, Hwang SK, Bajpai VK, Huh YS, Han YK. Current demands for food-approved liposome nanoparticles in food and safety sector. Frontiers in Microbiology. 2017;**8**:1-14. DOI: 10.3389/fmicb.2017.02398

[77] Machado AR, Assis LM, Costa JAV, Badiale-Furlong E, Motta AS, Micheletto YMS, et al. Application of sonication and mixing for nanoencapsulation of the cyanobacterium Spirulina platensis in liposomes. International Food Research Journal. 2014;**21**:2201-2206

Chapter 4

Dissipative Particle Dynamics Simulations of Self-Assemblies of Liposomes for Drug Delivery Applications

Ketzasmin Armando Terrón-Mejía, Inocencio Higuera-Ciapara, Evelin Martínez-Benavidez, Javier Hernández and Roberto López-Rendón

Abstract

Liposomes are essential components in the development of functional materials for drug delivery; this is mainly due to its ability to self-associate spontaneously and form bilayer vesicles. In these potential applications, knowing the size of self-assembled liposomes is essential for optimal performance; however, this process still has many unanswered questions. Conventional experimental techniques to study self-assemblies of liposome nanoparticles still have a great challenge. Computational simulations emerge as an alternative to understand the role of thermodynamic properties responsible for the self-assembly, particularly when they are unreachable experimentally because of limited time and length resolutions. In this chapter, we present the advantages and disadvantages of dissipative particle dynamic method to explore the functioning of liposome self-assembly in the transport of drugs.

Keywords: dissipative particle dynamics, liposomes self-assemblies, drug delivery, capsaicin, chitosan polymers

1. Introduction

The scenario of the modern era of interdisciplinary research between the natural sciences, including mathematics and theoretical physics, biophysics, chemistry, colloid science, and biochemistry, among others, along with advances in computational sciences, makes possible the manipulation of matter on a molecular scale through nanotechnology and bionanotechnology. This interdisciplinary research is able to study nanosystems made of phospholipids can be self-assembled spontaneously. One of the most studied cases in recent years is the self-assembly of liposomes.

The liposomes, vesicular systems based on lipids, in the last decades have increased their potential of their applications in various scientific disciplines and extensively been used as carriers for numerous molecules in cosmetic and pharmaceutical industries and medical and agricultural fields due to their high capacity and efficiency in drug delivery [1, 2]. For purposes in drug delivery systems, the liposomes are suitable candidates, due to their multiple properties that make them unique in these processes, such as biocompatibility, biodegradability, low toxicity, ability to modify the pharmacokinetic profile of the drug loaded, etc. [3]. In this way, liposomes have been successfully applied in the delivery of nucleic acid systems [4], in the intestinal lymphatic delivery [5], for enhanced oral delivery [6]. Allen and Cullis documented that these advances have led to numerous clinical trials in such diverse areas as the delivery of anticancer, antifungal, and antibiotic drugs; the delivery of gene medicines; and the delivery of anesthetics and anti-inflammatory drugs [7]. Despite the innumerable applications and advances of the drug delivery of liposomes, it presents enormous challenges still. For instance, He et al. argued that by modulating the compositions of the lipid bilayers and adding polymers or ligands, both the stability and permeability of liposomes can be greatly improved for oral drug delivery [8]. In order to achieve direct, effective, and selective delivery of the liposome itself, a deeper understanding of the mechanisms for interaction between the water/ligands/polymers and the liposome is required [9]. It has been reported in different sources that the mechanism of liposome formation is based on the unfavorable interactions occurring between amphiphilic compounds and water molecules, where the polar head groups of phospholipids are subjected to the aqueous phases of the inner, the hydrophobic hydrocarbon tails are associated with the bilayer, and spherical core shell structures are formed [10–12].

In addition to a large number of experiments that have been dedicated to the study of the mechanism of liposome formation, computational studies have made important contributions to the subject. These methodologies have been shown to be very effective in knowing in a very specific way the role played by molecular interactions in complex systems such as liposome self-assembly. Nowadays, it is increasingly common for research protocols to include a computational analysis prior to carrying out the experimental stage. Different techniques of computer modeling can facilitate quantitative understanding of experimental observations and secure fundamental interpretation of underlying phenomena [13]. For instance, Thota and Jiang in an interesting review documented that with the use of simulation methods, it is possible to elucidate the mechanisms of drug loading/release, which are indispensable in the rational screening and designing new amphiphiles for high-efficacy drug delivery [14].

Most of the computational studies for drug delivery use molecular dynamics (MD) or Monte Carlo (MC) simulations to obtain structural and dynamic information. In the case of MD, for example, the study of interactions is carried out considering all atoms explicitly commonly. These interactions are evaluated by analytical functions that form the so-called force fields. The level of precision of a simulation with MD relies on the force field used. There are traditional force fields such as CHARMM [15], AMBER [16], and OPLS [17] used to study the self-assembly of complex systems such as liposomes. Due to the scheme of evaluation of the interactions, which is through both the intra- and intermolecular forces, the MD simulations are slow and computationally expensive when the number of atoms is very large. This limitation makes it almost impossible to reach the time scales (of the order of the microseconds and beyond) where the thermodynamic observables have greater physical significance. For this reason, this technique turns out to be little viable to study the systems planned in this work, although it is important to point out that diverse groups use it with very successful results.

Modeling technique alternatives to MD have been developed with the aim of overcoming the bottleneck when it is required to study systems composed of a large number of atoms, as well as the issue of length time scales. One of those techniques is the dissipative particle dynamics (DPD). The main philosophy of the DPD model is based on the so-called coarse-grained (CG) models. These models allow to diminish considerably the number of atoms of a given system, grouping a set of atoms, which can be a functional chemical group, into a single interaction site called a *bead*, always trying to maintain the physics of the correct model. This size reduction facilitates larger simulations of larger ones with larger systems, allowing a more efficient exploration of phase space, but losing an atomistic description. Despite the success of the DPD technique, very few groups have used it to obtain detailed information on various aspects of self-assembly of liposomes. Goicochea et al. used DPD for understanding the behavior of biopolymer brushes coating drug-carrying liposomes in an aqueous environment [18]. Shen et al. with the help of DPD simulations, they explored how the tethered PEG polymers will affect the membrane wrapping process of PEGylated liposomes during endocytosis [19]. Yamamoto et al. used DPD simulations to study the spontaneous vesicle formation of amphiphilic molecules in aqueous solution [20].

Another great advantage of the modeling methods such as DPD is to be able to incorporate their algorithms into powerful supercomputing technologies such as graphic processing units (GPUs). The combination of hardware and software with DPD has allowed us to explore systems at time scales, which were unattainable a few years ago. For instance, Wu et al. [21] presented a GPU-accelerated DPD with parallel cell-list updating. On the other hand, Nguyen et al. [22] achieve a speedup of up to 9.5x in the DPD simulations in the well-known LAMMPS code using GPUs. Phillips et al. [23] also achieve important speedup of DPD using GPUs but the HOOMD-blue code. Other important implementations of DPD in GPUs can be found in Refs [24, 25]. Our group has been a pioneer in the use and application of DPD methodologies using GPUs. We have developed a simulation code called SIMES [26] (which is an acronym for simulation at mesoscopic time scale), a production DPD software designed and implemented specifically to run on GPUs. SIMES can simulate a broad spectrum of complex systems, including the self-assemblies of liposomes. Readers interested in going deeper into this topic can consult our contributions in the following references [27-30].

Recently, in a previous work of our group, we have studied the stability of structures and the efficiency of the encapsulation of capsaicin, as well as the internal and superficial distribution of capsaicin and chitosan inside of a nanoliposome, constituted by lecithin and coated with a shell of chitosan by DPD simulations [31]. We found that thermodynamic properties, such as the potential mean force, show that the interaction between capsaicin and chitosan polymers is very weak compared to that with lecithin. Besides, an association between capsaicin and chitosan in presence of lecithin is not likely to occur. In this work, we managed to combine the power of the DPD method with the power of the GPUs. The simulations reported have been the longest so far by regarding the encapsulation of capsaicin by liposomal nanoparticles.

Continuing with our research computational about self-assembly of molecular components for the design of bionanomaterials focused on drug delivery applications, the objective of this chapter is to show the advantages offered by the DPD technique to study the self-assembly mechanism of liposomes. As a specific case study, we show the formation of liposomes loaded with capsaicin. Capsaicin, a lipophilic drug, is the pungent vanilloid compound in spicy chili peppers. We chose capsaicin because it is approved as a drug for the treatment of chronic pains [32]. It is also an excellent candidate to be transported by natural polymers such as chitosan [33]. In addition, it is widely recognized in the treatment of urological disorders and the control of satiety and obesity [34], among other important applications.

2. Dissipative particle dynamic foundations

2.1 Theoretical formulation

The simulation of DPD is one of the most viable methods to study thermodynamic and structural properties of biophysical processes at mesoscopic time and length scales, unlike the simulation with MD, where insights at microscopic time and length scales can achieved. This great advantage of DPD simulation is that the DPD models are built under the coarse-grained scheme, which is obtained from the description of all-atom of a given system. This coarse-graining parameter plays a vital role and has significant impact on the speed of simulation [35]. This feature makes the DPD simulations cheap and very fast computationally speaking.

The theoretical foundations of the DPD method are documented elsewhere. Readers interested can consult the following references [36-40] for more specific details. In this chapter, we will mention in a very summarized way the main mathematical aspects that govern the equations of this methodology. The DPD was introduced by Hoogerbrugge and Koelman in 1992 [41] for simulations of hydrodynamic phenomena. It was subsequently modified by Español and Warren in 1995 [42] who introduced the fluctuation-dissipation theorem in the original algorithm to add the frictional and stochastic forces. Therefore, a coupling between the statistical mechanic laws of the beads and Gibbs canonical ensemble is given, favoring the calculation of thermodynamic properties at longer time and length scales. Similar to molecular dynamics, the time evolution of each DPD particle can be calculated by Newton's second law:

$$\frac{d\mathbf{r}_i}{dt} = \mathbf{v}_i, \frac{d\mathbf{p}_i}{dt} = \sum_{j \neq i} \mathbf{F}_{ij} \tag{1}$$

where $\mathbf{r}_i$, $\mathbf{v}_i$, and $\mathbf{p}_i$ are respectively the position, velocity, and momentum vectors of particle i, and $\mathbf{F}_{ij}$ is the total interparticle force exerted on particle i by particle j. Specifically, $\mathbf{F}_{ij} = \mathbf{F}_{ij}^C + \mathbf{F}_{ij}^R + \mathbf{F}_{ij}^R$, where a purely repulsive conservative force $\mathbf{F}_{ij}^C$, a dissipative or frictional force $\mathbf{F}_{ij}^R$, which represents the effects of viscosity and slows down the particle motion with respect to each other, and a random (stochastic) force $\mathbf{F}_{ij}^R$, which represents the thermal or vibrational energy of system, is summed to obtain the total force, and these force components can be individually written as

$$\begin{aligned}
\mathbf{F}_{ij}^C &= \begin{cases} a_{ij}(1 - r_{ij}/r_c)r_{ij} & (r_{ij} < r_c) \\ 0 & (r_{ij} \geq r_c) \end{cases}, \\
\mathbf{F}_{ij}^D &= -\gamma w^D(r_{ij})\left[\mathbf{v}_{ij} \cdot \mathbf{e}_{ij}\right]\mathbf{e}_{ij} \\
\mathbf{F}_{ij}^R &= -\sigma w^R(r_{ij})\theta_{ij}\mathbf{e}_{ij}
\end{aligned} \tag{2}$$

where $\mathbf{e}_{ij} = \mathbf{r}_{ij}/\mathbf{r}_{ij}$, $\mathbf{r}_{ij} = \mathbf{r}_i - \mathbf{r}_j$, $\mathbf{r}_{ij} = |\mathbf{r}_i - \mathbf{r}_j|$, and $\mathbf{v}_{ij} = |\mathbf{v}_i - \mathbf{v}_j|$. The parameters $w^D(r_{ij})$ and $w^R(r_{ij})$ are the distance-dependent weight functions for the dissipative force and random force, respectively. The symbols a_{ij}, γ, and σ determine the strength of conservative, random, and dissipative forces, respectively. Moreover, in order to generate a correct equilibrium Gibbs-Boltzmann distribution, the dissipative and random forces have to satisfy the following relations:

$$w^D(r_{ij}) = \left[w^R(r_{ij})\right]^2, \sigma^2 = 2\gamma k_B T \tag{3}$$

where k_B is the Boltzmann constant. According to Groot and Warren [43], we choose a simple form of $w^D(r_{ij})$ and $w^R(r_{ij})$ as follows:

$$w^D(r_{ij}) = \left[w^R(r_{ij})\right]^2 = \begin{cases} (1 - r_{ij}/r_c)^2 & (r_{ij} < r_c) \\ 0 & (r_{ij} \geq r_c) \end{cases} \tag{4}$$

Technically, DPD simulation differs from MD simulation in two respects. First, the conservative pairwise forces between DPD particles are soft-repulsive, which makes it possible to extend the simulations to longer time scales. Second, the scheme to maintain constant temperature in DPD is implemented in terms of dissipative as well as random pairwise forces such that the momentum is locally conserved, which results in the emergence of hydrodynamic flow effects on the macroscopic scale.

2.2 Mesoscopic model of liposomes

From the molecular point of view, a liposome, which is usually amphiphilic molecules, contains polar heads and hydrophobic hydrocarbon tails. This molecular structure is ideal to build a DPD model, since the polar part can be grouped in a single bead, while the hydrophobic part can be grouped in another bead. The liposome presented in this work is constituted by lecithin, which has two long hydrocarbon chains, is a major component of lipid bilayers of cell membranes and is a natural and biological amphiphile. Lecithin is a natural lipid mixture of phospholipids used frequently for the preparation of various nanosystem delivery vehicles, such as liposomes [9, 10]. According to the molecular structure of lecithin, this can be separated into three different beads that correspond to the head, neck, and tail groups, respectively. A schematic illustration of the fundamental self-assembly of the structure of a liposome adopted in this chapter is shown in **Figure 1**. In **Figure 1A**, a detailed molecular description at all-atom level for the lipid is shown; a mapping in the bead representation of the atomistic structure is also shown. **Figure 1B** is depicting the coarse-grained model of lecithin used in this work with its three fundamental parts. A snapshot of configurational structure of the liposome constructed based on lecithin is shown in **Figure 1C**. From this representation, the yellow spheres represent the hydrophobic part of the lecithin, while the red spheres represent the hydrophilic part.

On the other hand, we show the coarse-grained model of capsaicin, our target as case of study in drug delivery applications. The molecular structure of capsaicin is also ideal for building a DPD model. Similar to lecithin, the capsaicin can be grouped into three groups or beads, i.e., one bead for the head, neck, and tail groups, respectively. **Figure 1A** shows the coarse-grained model for capsaicin.

Finally, in **Figure 1A**, we show the coarse-grained model of water. This model in DPD simulation is already well known; for example, a bead DPD can group up to three water molecules. This model allows to conserve the thermodynamic properties of water.

2.3 Computational details of mesoscopic simulations

Mesoscopic simulations were performed to study the self-assembling of liposomes. The capsaicin was used as an example of drug delivery applications. For this end, the simulations presented in this work were made under canonical ensemble,

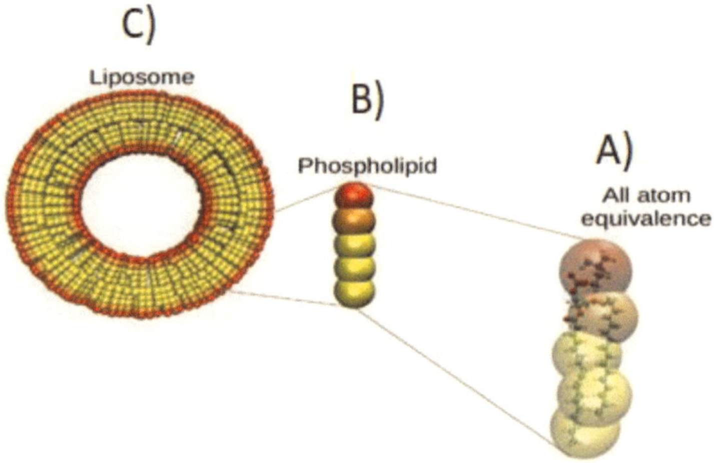

Figure 1.
Schematic illustration of computational models used for self-assembly of liposomes, from atomistic to mesoscopic representation. (A) Example of a lipid in a representation of all-atom. The computational model of coarse grain is constructed by grouping certain functional groups that form the fundamental molecular structure. In this case, the 35 atoms of the molecular structure of the phospholipid are grouped into only five sites or pseudo-atoms of coarse-grained type. (B) Example of our coarse-grained model of a lipid. The red spheres represent the hydrophobic group, while yellow spheres represent the hydrophilic group. (C) A representation of a liposome showing the inner of aqueous nucleus.

	L1	L2	L3	W
L1	78.0	80.0	95.0	89.0
L2	80.0	78.0	86.0	93.0
L3	95.0	86.0	78.0	101.0
W	89.0	93.0	101.0	78.0

Table 1.
Matrix interaction a_{ij} according to Eq. (2).

where N (the total particle number), V (volume), and T (temperature) are kept constant. For this type of simulations at constant temperature, some parameters must be specified. In this work, in all the simulations, we use dimensionless units or DPD units, i.e., the temperature and density were of T = 1.0, ρ = 3.0, respectively, the mass of all particles is $m = 1.0$. The system studied (liposome in aqueous solution) is constituted by 1000, 2000, 3000, 4000, and 5000 molecules of lecithin; each lecithin is constituted by three beads as shown in **Figure 1**; the remaining particles form part of solvent or water molecules. A total of 200,000 DPD particles were simulated. The simulation cell has a length of 40 DPD units. In DPD simulations, a key factor to carry them out is the specification of the parameters of interaction between the different components that make up a given system; this is due to the presence of the conservative forces in Eq. (3). The DPD interaction matrix for each type of bead used in this chapter is shown in **Table 1**.

The nomenclature used in this table goes as follows: the symbols *L1*, *L2*, and *L3* represent the head, neck, and tail beads of lecithin, while that *W* symbol represents the bead corresponding the beads of water.

On the other hand, the parameters of intramolecular interactions in lecithin are $r_0 = 0.7$ and $k_r = 100.0$ for the bonds between two particles [44], while for angles between three particles, it is taken $\theta_0 = 170.0$ and $k_\theta = 50.0$. Finally, the cutoff radius is taken as $r_c = 1.0$; the parameters σ and γ of the dissipative and random forces are equal to 3.0 and 4.5, respectively, parameters that appear in Eq. (2). The integration time step is $\Delta t = 0.03$, and the simulations are running over the 100 blocks of 10,000 steps every block, having a total of 1×10^6 steps, to reach a total of 4.8 μs. This simulation time is long enough to observe the self-assembly of the liposomes. With a traditional MD, the use of large supercomputing clusters is required to obtain this time scale. The equations of motion are solved using the velocity Verlet algorithm adapted to DPD [45]. The presence of electric charges was not considered in these calculations.

3. Drug delivery applications through numerical experiments

For the study of self-assemblies of liposomes for drug delivery applications, three different processes are presented. First, we analyze lipid self-assembly. Second, self-assembly of capsaicin is presented. Third, we end with the analysis of self-assembly of capsaicin by lecithin. The formation of nanoliposomes was explored by varying the concentration of both lecithin and capsaicin molecules separately. These simulations were analyzed through its density maps. This property is very useful for refining structures or adjusting molecular models; besides, the density maps allow us to spatially inspect the average location of atoms of interest during a simulation [46].

3.1 Exploring self-assemblies of lipids

To explore the self-assemblies of lipids, the density maps of five different processes were obtained. In these processes, the effect of lipid concentration was explored. **Figure 2** shows the snapshots of each of these processes. We observe that in **Figure 2A**, with a concentration of 1000 lipid molecules, the self-association of lipids is practically negligible. Only a slight formation of small clusters is perceived, similar to the density maps of very diluted solutions. By increasing the concentration of lipids to double, that is, to 2000 molecules as can be seen in **Figure 2B**, we observe that the association increases with respect to the previous case of 1000 lipid molecules. However, it is not yet a sufficient concentration to detect any indication of the process of self-assembly of lipids. **Figure 2C** shows the case for a concentration of 3000 lipid molecules. In this system, the formation of very faint regions is observed. The presence of these very light red dots is an indication that the lipids begin to self-assemble. To verify that the lipid molecules are indeed self-assembling, a simulation with 4000 lipid molecules was carried out. The result is shown in **Figure 2D**. As we expected, the formation of the red dots increases, and they begin

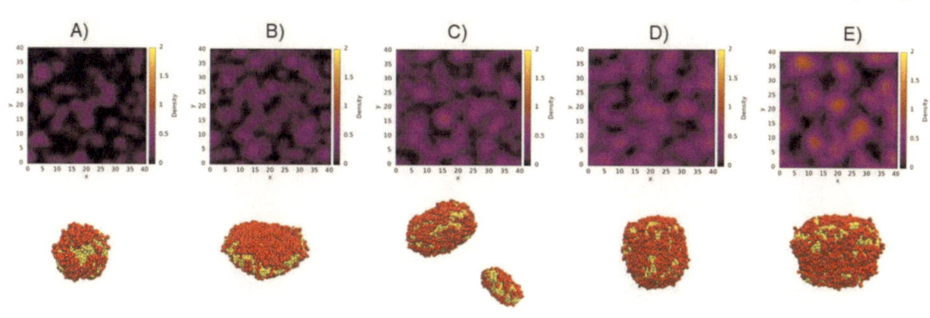

Figure 2.
Density maps for exploring the self-assembly of lipids. (A) 1000 lipid molecules, (B) 2000 lipid molecules, (C) 3000 lipid molecules, (D) 4000 lipid molecules, and (E) 5000 lipid molecules. Below each density map, a snapshot of the final configuration is displayed using the same color code as ***Figure 1****. The solvent molecules are not shown for clarity purposes.*

more notoriously. To observe the stability in the formation of lipids, a simulation with 5000 molecules was carried out. At this concentration, the lipid association is already clearly observed. A picture of this system is seen in **Figure 2E**. The red regions show that the self-assembly process has already taken place. These are the optimal conditions in terms of concentration for the process of self-assembly of lipids to be carried out. For higher concentrations of lipids, greater than 5000 molecules, we will find a saturation of the system, which is an amount sufficient for the formation of a nanoliposome. An animation video of each of these systems can be found in the supplementary material section of this work.

3.2 Exploring self-assemblies of capsaicin

We conducted an exploration to study the self-assembly of capsaicin. For this purpose, we carried out four separate numerical experiments. The results of these simulations are found in **Figure 3**, analyzed through their density maps. These maps show regions of high and low density basically. The regions of low density are the ones that predominate the colors black and purple, while the regions of high density are those that are identified in lighter shades such as red or yellow. **Figure 3A** shows the density map for a system composed of 250 capsaicin molecules. Due to the low concentration of capsaicin molecules, the black and purple regions predominate, which are located in a density range between 0.02 and 0.04. In the second numerical experiment, we doubled the concentration of capsaicin molecules. **Figure 3B** shows the density maps for this system composed of 500 capsaicin molecules. The presence of regions in all reddish is notorious, which oscillates between 0.08 and 0.12 in density. However, purple regions are still observed, which indicate that this

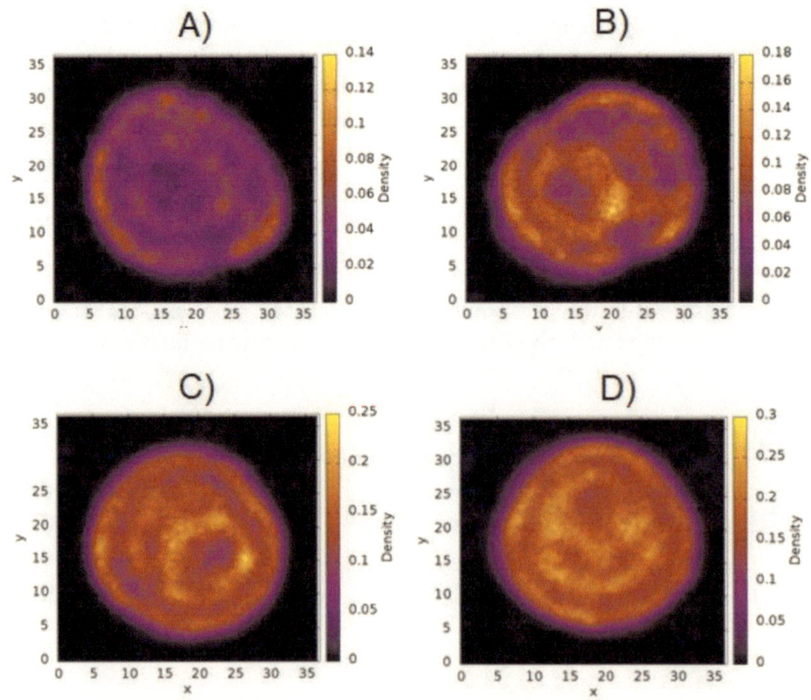

Figure 3.
Density maps for exploring the self-assembly of capsaicin. These snapshots correspond to equilibrated systems composed of (A) 250 capsaicin molecules, (B) 500 capsaicin molecules, (C) 750 capsaicin molecules, and (D) 1000 capsaicin molecules.

concentration of capsaicin molecules is not yet enough to have a homogenous system. **Figure 3C** shows the density map for a system composed of 750 capsaicin molecules. Unlike the density map of **Figure 3B**, the presence of purple areas is already minimal, while the presence of high-density areas is more consolidated. A higher density of capsaicin molecules is necessary to completely eliminate the purple areas. This can be seen in **Figure 3D**, which shows the density map for a concentration of 1000 capsaicin molecules. Note in this case how the purple areas have completely disappeared. The entire density map is in a homogenous system with a density that ranges between 0.2 and 0.25. This concentration of capsaicin molecules is optimal for use in the process of formation of nanoliposomes.

3.3 Exploring self-assembly of capsaicin by lecithin nanoliposomes

We have documented that nanoliposomes are effective for the drug delivery process. A strategic requirement for the supply of medicines to a specific site has driven the development of active drug targeting techniques [47]. In this way, to take advantage of the effective potential of the liposome technology and the impact on its formulations, it is necessary to know the optimal conditions in terms of concentration of each of the components that are required for an optimal application of any given drug. To explore the self-assembly of capsaicin by means of lecithin liposomes, we conducted four numerical experiments. The number of capsaicin molecules used in these four experiments was 250, 500, 750, and 1000 corresponding to a concentration of 30, 61, 92, and 123 mM, respectively, and the concentration of lecithin remained constant with 5000 molecules in the four cases. **Figure 4** shows the density maps obtained for these four cases. **Figure 4A–D** shows the density maps for a system loaded with 250, 500, 750, and 1000 capsaicin molecules. For all four we observed a homogeneous formation of the liposome. The

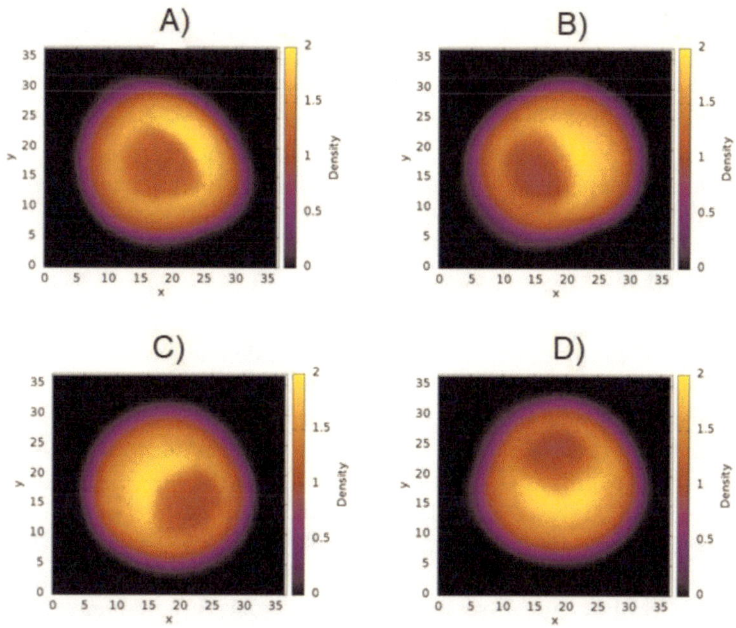

Figure 4.
Density maps for exploring self-assembly of capsaicin by lecithin nanoliposomes. These snapshots correspond to equilibrated systems composed of (A) 250 capsaicin molecules, (B) 500 capsaicin molecules, (C) 750 capsaicin molecules, and (D) 1000 capsaicin molecules.

Concentration of capsaicin (Mn)	Size (nm)	± (nm)	EE (%)	± (%)
30	17.95	0.46	96.27	0.69
61	18.01	0.61	96.93	0.36
92	17.90	0.21	96.81	0.35
123	17.97	0.16	96.94	0.25

Table 2.
Mean size of the nanoliposome and encapsulation efficiency (EE) as a function of concentration of capsaicin.

high-density regions are representing the region formed by the lipid membrane, which in this case is presented as the region with the highest concentration (in yellow) in the four cases. In addition, in the four cases, they present a redder region in the center of the map; this indicates that the density of the lipids is lower, as expected. Density maps also show that the nanoliposome is stable for 4.8 μs of simulation, since the lecithin molecules do not spread throughout the simulation box nor do they collapse in the aqueous core to form a micelle; this was possible thanks to the previous analysis where we found the optimal concentration of lecithin molecules in the self-assembly process.

In our previous study [31], we observed the formation of a nanoliposome based on lecithin was carried out with approximately 7000 molecules of lecithin, but in that study, the presence of natural polymers such as chitosan was included, a component that is not present in these simulations.

Derived from the results obtained in the density maps of **Figure 4**, we proceeded to make an analysis on the encapsulation efficiency (EE). It has been reported that both size and EE are key variables in the preparation of liposomes [47]. However, these variables are difficult to control at the laboratory level even with sophisticated experimental equipment. Computational tools such as DPD are ideal for guiding experiments. The size and EE for the system analyzed in this work are show in **Table 2**. EE is obtained from the following equation: $EE = ((CapsT - CapsF)/CapsT) \times 100$, where CapsT is the total concentration of capsaicin in the system and CapsF is the free capsaicin in solution. The percentage of EE for the capsaicin obtained in this work was of 96%, which is very close to that reported experimentally of 92% [48], besides, is very similar to the obtained percentage in our previous study, where the presence of chitosan polymer was included [31].

4. Conclusions

In this chapter, simulations at the mesoscopic scale were used to systematically study the self-assembly of liposomes and their potential applications toward the delivery of drugs. The advantages offered by the mesoscopic methods such as DPD over other simulation methods such as MD were exposed; additionally, the acceleration that the DPD simulations have achieved through the new GPUs technologies was also covered. With these technologies, we show simulations in the microsecond scale with a system composed of 150,000 particles in the process of self-assembly of liposomes. We show the optimal conditions for the formation of liposomes in function of the lecithin concentration to achieve a good capsaicin encapsulation. Our percentage of EE obtained is quite acceptable with respect to experimental measurements and simulation studies of similar systems.

Acknowledgements

This work was supported by CONACYT [Project FON.INST./44/2016] and SIEA-UAEM [Project 4728/2019CIB]. All simulations reported in this work were performed at the Supercomputer OLINKA, located at the Laboratory of Molecular Bioengineering at Multiscale (LMBM), at the Universidad Autónoma del Estado de México.

Conflict of interest

The authors declare that they have no competing interests.

Author details

Ketzasmin Armando Terrón-Mejía[1], Inocencio Higuera-Ciapara[2], Evelin Martínez-Benavidez[2], Javier Hernández[3] and Roberto López-Rendón[1]*

1 Laboratorio de Bioingeniería Molecular a Multiescala, Facultad de Ciencias, Universidad Autónoma del Estado de México, Toluca, Estado de México, México

2 Centro de Investigación y Asistencia en Tecnología y Diseño del Estado de Jalisco, A.C., Guadalajara, Jalisco, México

3 Unidad de Servicios de Apoyo en Resolución Analítica, Universidad Veracruzana, Xalapa, Veracruz, México

*Address all correspondence to: roberto.lopez.rendon@gmail.com

References

[1] Liu W, Ye A, Han F, Han J. Advances and challenges in liposome digestion: Surface interaction, biological fate, and GIT modeling. Advances in Colloid and Interface Science. 2019;**263**:52-67. DOI: 10.1016/j.cis.2018.11.007

[2] Akbarzadeh A, Rezaei-Sadabady R, Davaran S, Joo SW, Zarghami N, Hanifehpour Y, et al. Liposome: Classification, preparation, and applications. Nanoscale Research Letters. 2013;**8**:1-9. http://www.nanoscalereslett.com/content/8/1/102

[3] Sawant RR, Torchilin VP. Challenges in development of targeted liposomal therapeutics. The AAPS Journal. 2012; **14**:303-315. DOI: 10.1208/s12248-012-9330-0

[4] Shim G, Mi-Gyeong Kim M-G, Park JY, Oh Y-K. Application of cationic liposomes for delivery of nucleic acids. Asian Journal of Pharmaceutical Sciences. 2013;**8**:72-80. DOI: 10.1016/j.ajps.2013.07.009

[5] Kim H, Yeongseok Kim Y, Lee J. Liposomal formulations for enhanced lymphatic drug delivery. Asian Journal of Pharmaceutical Sciences. 2013;**8**:96-103. DOI: 10.1016/j.ajps.2013.07.012

[6] He H, Lu Y, Qi J, Zhao W, Dong X, Wu W. Biomimetic thiamine- and niacin-decorated liposomes for enhanced oral delivery of insulin. Acta Pharmaceutica Sinica B. 2018;**8**:97-105. DOI: 10.1016/j.apsb.2017.11.007

[7] Allen TM, Cullis PR. Liposomal drug delivery systems: From concept to clinical applications. Advanced Drug Delivery Reviews. 2013;**65**:36-48. DOI: 10.1016/j.addr.2012.09.037

[8] He H, Lu Y, Qi J, Zhu Q, Chen Z, Wu W. Adapting liposomes for oral drug delivery. Acta Pharmaceutica Sinica B. 2019;**9**:36-48. DOI: 10.1016/j.apsb.2018.06.005

[9] Daraee H, Etemadi A, Kouhi M, Alimirzalu S, Akbarzadeh A. Application of liposomes in medicine and drug delivery. Artificial Cells, Nanomedicine, and Biotechnology. 2014:1-11. DOI: 10.3109/21691401.2014.953633

[10] Fathi M, Mozafari MR, Mohebbia M. Nanoencapsulation of food ingredients using lipid based delivery systems. Trends in Food Science & Technology. 2012;**23**:13-27. DOI: 10.1016/j.tifs.2011.08.003

[11] Jesorka A, Orwar O. Liposomes: Technologies and analytical applications. Annual Review of Analytical Chemistry. 2008;**1**:801-832. DOI: 10.1146/annurev.anchem.1.031207.112747

[12] Sercombe L, Veerati T, Moheimani F, Wu SY, Sood AK, Hua S. Advances and challenges of liposome assisted drug delivery. Frontiers in Pharmacology. 2015;**6**:286. DOI: 10.3389/fphar.2015.00286

[13] Decuzzi P. Facilitating the clinical integration of nanomedicines: The roles of theoretical and computational scientists. ACS Nano. 2016;**10**:8133-8138. DOI: 10.1021/acsnano.6b05536

[14] Thota N, Jiang J. Computational amphiphilic materials for drug delivery. Frontiers in Materials. 2015;**2**:1-15. DOI: 10.3389/fmats.2015.00064

[15] Klauda JB, Venable RM, MacKerell AD Jr, Pastor RW. Considerations for lipid force field development, computational modeling of membrane bilayers. Current Topics in Membranes. 2008;**60**:1-48. DOI: 10.1016/S1063-5823(08)00001-X

[16] Siu SWI, Vácha R, Jungwirth P, Böckmann RA. Biomolecular simulations of membranes: Physical properties from different force fields. The Journal of Chemical Physics. 2008; **128**:125103. DOI: 10.1063/1.2897760

[17] Jorgensen WL, Tirado-Rives J. The OPLS [optimized potentials for liquid simulations] potential functions for proteins, energy minimizations for crystals of cyclic peptides and crambin. Journal of the American Chemical Society. 1988;**110**:1657-1666. DOI: 10.1021/ja00214a001

[18] Goicochea AG, Mayoral E, Klapp J, Pastorino C. Nanotribology of biopolymer brushes in aqueous solution using dissipative particle dynamics simulations: An application to PEG covered liposomes in a theta solvent. Soft Matter. 2014;**10**:166-174. DOI: 10.1039/c3sm52486h

[19] Shen Z, Ye H, Kröger M, Li Y. Aggregation of polyethylene glycol polymers suppresses receptor-mediated endocytosis of PEGylated liposomes. Nanoscale. 2018;**10**:4545-4560. DOI: 10.1039/c7nr09011k

[20] Yamamoto S, Maruyama Y, Hyodo S-A. Dissipative particle dynamics study of spontaneous vesicle formation of amphiphilic molecules. The Journal of Chemical Physics. 2002;**116**:5842-5849. DOI: 10.1063/1.1456031

[21] Wu H, Xu J, Zhang S, Wen H. GPU accelerated dissipative particle dynamics with parallel cell-list updating. IEIT Journal of Adaptive & Dynamic Computing. 2011;(2):26-32. DOI: 10.5813/www.ieit-web.org/IJADC/2011.2.4

[22] Nguyen TD, Phillips CL, Anderson JA, Glotzer SC. Rigid body constraints realized in massively-parallel molecular dynamics on graphics processing units. Computer Physics Communications. 2011;**182**:2307-2313. DOI: 10.1016/j.cpc.2011.06.005

[23] Phillips CL, Anderson JA, Glotzer SC. Pseudo-random number generation for Brownian dynamics and dissipative particle dynamics simulations on GPU devices. Journal of Computational Physics. 2011;**230**:7191-7201. DOI: 10.1016/j.jcp.2011.05.021

[24] Blumers AL, Tang Y-H, Li Z, Li X, Karniadakis GE. GPU-accelerated red blood cells simulations with transport dissipative particle dynamics. Computer Physics Communications. 2017;**217**: 171-179. DOI: 10.1016/j.cpc.2017.03.016

[25] Tang Y-H, Karniadakis GE. Accelerating dissipative particle dynamics simulations on GPUs: Algorithms, numerics and applications. Computer Physics Communications. 2014;**185**:2809-2822. DOI: 10.1016/j.cpc.2014.06.015

[26] Terrón-Mejía KA, López-Rendón R, Goicochea AG. SIMES: SImulation at MESoscopic Scale—Accelerating dissipative particle dynamics simulations on graphical processing units. In: The 3rd International Electronic and Flipped Conference on Entropy and Its Applications (ECEA 2016), 1–10 November 2016; Sciforum Electronic Conference Series, Vol. 3; 2016. DOI: 10.3390/ecea-3-A001

[27] Terrón-Mejía KA, López-Rendón R, Goicochea AG. Mesoscopic modeling of structural and thermodynamic properties of fluids confined by rough surfaces. Physical Chemistry Chemical Physics. 2015;**17**:26403-26416. DOI: 10.1039/c5cp03823e

[28] Terrón-Mejía KA, López-Rendón R, Goicochea AG. In: Gitler I, Klapp J, editors. ISUM 2015, CCIS 595A New Effective Potential for Colloidal Dispersions with Polymer Chains Emerging from Mesoscopic Scale Interactions. Switzerland: Springer

International Publishing; 2016, 2016. pp. 447-460. DOI: 10.1007/978-3-319-32243-8_31

[29] Terrón-Mejía KA, López-Rendón R, Goicochea AG. Electrostatics in dissipative particle dynamics using Ewald sums with point charges. Journal of Physics: Condensed Matter. 2016;**28**:425101. (10 pp). DOI: 10.1088/0953-8984/28/42/425101

[30] Terrón-Mejía KA, López-Rendón R, Goicochea AG. Desorption of hydrocarbon chains by association with ionic and nonionic surfactants under flow as a mechanism for enhanced oil recovery. Scientific Reports. 2017;**7**: 9586. DOI: 10.1038/s41598-017-09735-8

[31] Terrón-Mejía KA, Martínez-Benavidez E, Higuera-Ciapara I, Virués C, Hernández J, Domínguez Z, et al. Mesoscopic modeling of the encapsulation of capsaicin by lecithin/chitosan liposomal nanoparticles. Nanomaterials. 2018;**8**:425. DOI: 10.3390/nano8060425

[32] Bevan S, Szolcsányi J. Sensory neuron-specific actions of capsaicin: Mechanisms and applications. Trends in Pharmacological Sciences. 1990;**11**: 330-333. DOI: 10.1016/0165-6147(90)90237-3

[33] Higuera-Ciapara I, Virúes C, Jiménez-Chávez M, Martínez-Benavidez E, Hernández J, Domínguez Z, et al. ^{1}H NMR studies of molecular interaction of D-glucosamine and Nacetyl-D-glucosamine with capsaicin in aqueous and non-aqueous media. Carbohydrate Research. 2017;**452**:6-16. DOI: 10.1016/j.carres.2017.09.014

[34] Caterina MJ, Schumacher MA, Tominaga M, Rosen TA, Levine JD, Julius D. The capsaicin receptor: A heat-activated ion channel in the pain pathway. Nature. 1997;**389**:816-824. DOI: 10.1038/39807

[35] Backera JA, Lowe CP, Hoefsloot HCJ, Iedema PD. Combined length scales in dissipative particle dynamics. The Journal of Chemical Physics. 2005; **123**:114905. DOI: 10.1063/1.2013208

[36] Español P, Warren PB. Perspective: Dissipative particle dynamics. The Journal of Chemical Physics. 2017;**146**: 150901. DOI: 10.1063/1.4979514

[37] Moeendarbary E, Ng TY, Zangeneh M. Dissipative particle dynamics: Introduction, methodology and complex fluid applications—A review. International Journal of Applied Mechanics. 2009;**1**:737-763. DOI: 10.1142/S1758825109000381

[38] Li X, Li Z, Bian X, Deng M, Kim C, Tang Y-H, et al. Dissipative particle dynamics: An overview. In: Bhushan B, editor. Encyclopedia of Nanotechnology. Dordrecht: Springer

[39] Liu MR, Liu GR, Zhou LW, Chang JZ. Dissipative Particle Dynamics (DPD): An overview and recent developments. Archives of Computational Methods in Engineering. 2015;**22**:529-556. DOI: 10.1007/s11831-014-9124-x

[40] Guigas G, Morozova D, Weiss M. Exploring membrane and protein dynamics with dissipative particle dynamics. Advances in Protein Chemistry and Structural Biology. 2011; **85**:143-182. DOI: 10.1016/B978-0-12-386485-7.00004-1

[41] Hoogerbrugge PJ, Koelman JMVA. Simulating microscopic hydrodynamic phenomena with dissipative particle dynamics. Europhysics Letters. 1992;**19**: 155-160. DOI: 10.1209/0295-5075/19/3/001

[42] Español P, Warren P. Statistical mechanics of dissipative particle dynamics. Europhysics Letters. 1995;**30**:

191-196. DOI: 10.1209/0295-5075/30/4/001

[43] Groot RD, Warren PB. Dissipative particle dynamics: Bridging the gap between atomistic and mesoscopic simulation. The Journal of Chemical Physics. 1997;**107**:4423-4435. DOI: 10.1063/1.474784

[44] Goicochea AG, Romero-Bastida M, López-Rendón R. Dependence of thermodynamic properties of model systems on some dissipative particle dynamics parameters. Molecular Physics 2007, 105, 2375–2381.158. DOI: 10.1080/00268970701624679

[45] Vattulainen I, Karttunen M, Besold G, Polson JM. Integration schemes for dissipative particle dynamics simulations: From softly interacting systems towards hybrid models. The Journal of Chemical Physics. 2002;**116**: 3967-3979. DOI: 10.1063/1.1450554

[46] Briones R, Blau C, Kutzner C, de Groot BL, Aponte-Santamaría C. GROmaρs: A GROMACS-based toolset to analyze density maps derived from molecular dynamics simulations. Biophysical Journal. 2019;**116**:1-8. DOI: 10.1016/j.bpj.2018.11.3126

[47] Pattni BS, Chupin VV, Torchilin VP. New developments in liposomal drug delivery. Chemical Reviews. 2015;**115**: 10938-10966. DOI: 10.1021/acs.chemrev.5b00046

[48] Goycoolea FM, Milkova V. Electrokinetic behavior of chitosan adsorbed on o/w nanoemulsion droplets. Colloids and Surfaces A: Physicochemical and Engineering Aspects. 2017;**519**:205-211. DOI: 10.1016/j.colsurfa.2016.05.093

Chapter 5

Pharmaceutical Development of Liposomes Using the QbD Approach

Alina Porfire, Marcela Achim, Cristina Barbalata, Iulia Rus, Ioan Tomuta and Cecilia Cristea

Abstract

Quality by Design (QbD) is a systematic, risk-based approach to pharmaceutical product and manufacturing development, which uses quality-improving scientific methods upstream in the research, development, and design phases, in order to assure that quality and safety are designed into product at as early stage as possible. This work focuses on the state-of-the-art applications of the QbD principles in the development of liposomes. The QbD approach has recently been proposed as a useful tool to obtain higher-quality liposomal products, as their development is a challenging task, involving intricate formulation and manufacturing processes. Thus, the current strategies to define the relationship between the critical material attributes or process parameters and product critical quality attributes and to establish the design space are overviewed. Additionally, the current characterization methodologies are described, as part of the control strategy required within the QbD paradigm.

Keywords: liposomes, Quality by Design, critical quality attributes, design space, quality control

1. Introduction

In the recent years, the Quality by Design (QbD) concept has gained importance in drug development and drug manufacturing. QbD is recommended by the drug regulatory agencies (FDA—Food and Drug Administration and EMA—European Medicines Agency) to improve the quality of pharmaceutical products.

For the pharmaceutical products, quality is regarded as a mandatory topic and must be assured for all. Pharmaceutical products' quality is ensured through the ability to get the therapeutic benefit mentioned on the label and through the absence of contamination [1, 2]. Product quality refers to performance, robustness, trustworthiness, and has to be built into it [3]. To ensure the quality of a product, scientific approaches such as QbD can be implemented. The concept of QbD was first defined a few decades ago by Dr. Joseph M. Juran, a well-known pioneer of quality, and emphasizes the design of a product and manufacturing process to reach a certain predefined quality [1, 2, 4]. According to Dr. Juran, quality should be built into a product, and the way in which a product is designed is accountable for most

IntechOpen

issues related to its quality. Dr. Juran believed that the quality of a product could be planned and the most quality inadequacies originate from the way in which the quality of the product was planned [4, 5].

In the pharmaceutical field, there are several regulatory guidelines developed by ICH, FDA, and EMA, which offer the necessary information in understanding how the QbD concept may be implemented [2, 3, 6–8]. In ICH Q8—Pharmaceutical Development guideline, QbD is defined as "a systematic approach to development that begins with predefined objectives and emphasizes product and process understanding and process control, based on sound science and quality risk management" [9]. Such a scientific, risk-based, and pro-active approach from the development to manufacturing of a pharmaceutical product will provide the necessary knowledge and information to minimize the risk and ensure a predefinite quality of the product [10]. So, QbD provides the tool to understand the way in which the quality of a pharmaceutical product is influenced by formulation, input materials' characteristics, and process variables; therefore, the quality of the product can be ensured by controlling the formulation input materials and the manufacturing process key variables [2, 4, 11]. It involves designing a formulation and manufacturing process in such a way to obtain a pharmaceutical product with predetermined quality specifications [12]. QbD identifies characteristics that are critical to quality of the product, translates them into attributes that the drug should possess, and establishes how the critical formulation and process parameters can be varied to constantly produce a drug product with the desired characteristics [13, 14]. The goal of the QbD approach is in-depth understanding of the formulation and process variables, and of the relationship between them, in order to obtain a drug product with consistent desired characteristics [15, 16]. Practically, QbD helps establishing the critical quality attributes (CQAs) of the product and how the critical material attributes (CMAs) and critical process parameters (CPPs) can be modified to deliver a product with predetermined quality specifications [12].

A QbD approach includes several key steps, as follows: (1) defining the quality target product profile (QTPP); (2) performing a risk assessment, in order to identify which formulation, material, or process parameters can potentially influence the product's quality attributes (CQAs); (3) studying the impact of the formulation variables, material attributes, and process parameters on the critical quality attributes (CQAs) of the drug product, and finding which of them are critical material attributes (CMAs) and critical process parameters (CPPs); (4) establishing a design space that ensures desired product specifications; and (5) designing and implementing a control strategy in order to ensure a continuous improvement [9, 17, 18]. In order to study the relationship between the CMAs and CPPs and their impact on CQAs in a mathematical form, the Design of Experiments (DoE) strategy is used. This method also allows to establish the design space by running a minimum number of experiments [2, 11, 19]. The design space for liposome preparation may be established by implementing the QbD strategy as a systematic approach in liposome development, in order to improve the product quality, by understanding and controlling formulation, materials, and manufacturing variables. The strategy is recommended by the drug regulatory agencies for the development of better-quality products and may be used in liposomal drug product development.

2. Defining the QTPP for liposomal products and identification of the CQAs

The most important element in using QbD concept to assist formulation and process design is to predefine the desired final product quality profile [9, 14, 20].

QTPP elements		Target
Dosage form		Nanoformulation
Dosage design		For targeted delivery
Administration route		Parenteral
Quality attributes of the liposomal product	Biocompatibility	Lack of hemolytic activity
	Microbiological quality	Sterile
	Bacterial endotoxins	Free
	Physical attributes (particle size and morphology, viscosity, zeta potential, osmolarity, appearance) Drug identification Drug content *In vivo* stability Drug release Degradation products Residual solvents	Must meet the standards resulted from the specifications of similar approved products or from the current scientific research

Table 1.
QTPP of liposomal products.

According to the definition of ICH, the QTPP is "a prospective summary of the quality characteristics of a drug product that ideally will be achieved to ensure the desired quality, taking into account safety and efficiency of the drug product" [9]. In order to establish the QTPP, the following considerations must be taken into account: route of administration, dosage form, dosage strength, delivery system, attributes affecting pharmacokinetic characteristics, stability, sterility, and drug release appropriate for the intended final product [14, 20].

QTPP is established based on prior scientific knowledge and appropriate *in vivo* relevance. According to the current flux of literature, a QTPP for liposomal products is presented in **Table 1** along with the targets for each element [21].

When designing a nanoformulation, the efficiency of the final product will be directly related to particle size. Particle size is the most important factor influencing biodistribution, circulation half-life, and cellular uptake. Due to their small size and large surface area, pharmaceutical nanosystems show enhanced bioavailability and additional ability to cross the biological membranes. Furthermore, in cancer therapy, smart nanoparticles deliver the drug into the tumor tissue and avoid normal tissues and organs [22]. After intravenous administration, liposomes accumulate in tumors by a passive or active targeting, for determining higher therapeutic efficiency and less side effects [23]. As a parenteral dosage form, liposomal products must be sterile, pyrogen free, and well tolerated [24]. The lack of hemolytic activity is also mentioned as requirement for liposome products [25].

The second step in a QbD approach is the identification of CQAs. According to ICH Q8 definition, a CQA is a "physical, chemical, biological, or microbiological property or characteristic that should be within an appropriate limit, range, or distribution to ensure the desired product quality." They are either derived from the QTPP or defined based on regulatory requirements and review of the literature [9].

Based on these recommendations, for the liposomal products the following CQAs are usually identified: mean particle size, size distribution, and zeta potential (as physical CQAs); drug content (as chemical CQAs); *in vivo* stability and drug release (as biological CQAs); sterility (as microbiological CQA).

a. Generally, for all nanoparticles, the mean particle size and particle size distribution are major CQAs, which play an important role in determining their *in vivo* distribution, drug loading, drug release, and targeting ability.

The biodistribution of nanoparticles circulating in the blood stream is considerably influenced by particle size. Smaller particles, of 20–30 nm, are eliminated by kidneys, while larger particles, of up to 300 nm, are taken up by reticuloendothelial system (RES) [26]. By decreasing particle size below 100 nm, RES uptake can be reduced [27]. According to other studies, the ideal nanocarriers should have particle size ranging between 10 and 100 nm in order to evade the kidney filtration and capture by the liver [28]. For example, in tumor targeting, the particles need to be smaller than the cut-off of the fenestrations in the tumor neovasculature [29]. The literature reveals that for an efficient enhanced permeability and retention (EPR) effect, the particle size must be, generally, between 50 and 100 nm, depending on the tumor type, its environment, and its localization [28]. Particle size less than 200 nm is also beneficial for the sterility of the liposomal product, since this size allows aseptic preparation and sterile filtration of the final product [30].

Particle size also influences the kinetics of drug release. As particle size gets smaller, their surface area-to-volume ratio gets larger. This would imply that more of the drug is closer to the surface of the particle compared to a larger particle. Being at or near the surface would lead to a faster drug release [23].

Polydispersity index, reflecting particle size distribution of liposomal products, is a physical parameter for which the target is to have reduced values, indicating a good homogeneity of the dispersion. Generally, polydispersity index values below 0.5 are reported to be acceptable [31].

b. Zeta potential or particle charge is an important parameter in the evaluation of colloidal system's stability. Particles with a high negative or positive zeta potential value repel each other, indicating that the colloidal system is stable [32]. On the contrary, decreasing the zeta potential value to nearly neutral leads to liposomal aggregation [20]. The charge of nanosystems influences both systemic circulation time and the interactions with the target tissue. The presence of surface charge can alter the opsonization profile of the particles, its recognition by cells in the organs of the RES, and the overall plasma circulation profile [29]. Regarding the interaction with the target, it is known that cationic liposomes have affinity for negatively charged cancer cell membrane and higher selectivity than neutral or anionic liposomes. The intracellular uptake of cationic liposomes by tumor cells can be 14-fold higher than that of normal liposomes. Consequently, cationic liposomes enhance the safety of liposomal drug. The liposome charge can also influence drug loading, cationic liposomes exhibiting higher encapsulation efficiency for negatively charged drugs [32].

c. Depending on drug distribution in liposomal dispersion, the drug content of liposomes can be expressed in three ways: drug concentration (μg/mL); entrapment efficiency (EE, the amount of drug contained inside liposomes compared with total amount of drug; %); and drug loading (the amount of drug contained relative to the amount of the lipid used; drug-to-lipid ratio) [14, 20].

High encapsulation efficiency is very important for both manufacturers and patients. A higher percentage of drug encapsulation could reduce the manufacturing cost and increase drug concentration in the final formulation allowing greater flexibility in dosing. Depending on the pharmacokinetics, higher drug concentration can result in increased dosing intervals and hence improved patient compliance [30].

Drug retention inside liposomes is also very important. Since liposomes are intended to deliver the drug to the target site, there should be no drug leakage until cellular uptake [30].

d. In order to have a prolonged drug release and an efficient tumor accumulation by passive targeting, *in vivo* stability is essential. *In vivo* behavior of liposomes after intravenous administration is directly linked to their interaction with blood components, which depends on the hydrophobic/hydrophilic character of liposome surface. The more hydrophobic a nanoparticle is, the more likely it is to be cleared by phagocytosis, due to higher binding of blood proteins (opsonization) [22]. The improved surface properties are associated with the steric hindrance effect offered by hydrophilic polymers, which can prevent the surface-modified liposomes from being rapidly eliminated by RES ("stealth liposomes"). Hydrophilization of liposome surface prolongs circulation time (long-circulating liposomes), enhancing the therapeutic potential of the entrapped drug [33].

e. Drug release kinetics has important implication for the therapeutic activity of liposomes. The drug must be delivered to the disease site at a level within its therapeutic window, at a sufficient rate, for a sufficient period, to have optimal therapeutic activity [34]. Drug delivery to the right site is related to the identification and the interaction with the target cells. Cell surface or blood vessel surface of tumor tissues has a series of specific and overexpressed receptors, which are closely related with tumor process [33]. Targeting ligands attached on liposomes' surface bind the corresponding receptors and, as a result, the liposomes are internalized by endocytosis. These ligands can be monoclonal antibodies, fragments of antibodies, peptides, proteins, nucleic acids, carbohydrates, or small molecules [29].

3. The critical material attributes and critical process parameters: identification and linking to CQAs

All the material attributes (MAs) and/or process parameters (PPs) that can affect the desired critical quality attributes (CQAs) of the final/intermediate product are identified through risk assessment. After risk identification, risk analysis evaluates the impact of the identified MAs and PPs on the CQAs. Further, through risk evaluation, a qualitative or quantitative scale is used for risk estimation of each identified factor on the desired CQA [35]. After the risk evaluation of potential MAs and PPs, only few of them will become potentially critical for quality attributes of the final drug product. In this case, identified MAs will become critical material attributes (CMAs), which must possess certain characteristics or should be chosen in an appropriate range to assure CQAs of the intermediate/final drug product [21]. Critical process parameters (CPPs) are those PPs that should be monitored and controlled in order to obtain the desired CQAs of the intermediate/final drug product [9]. Many tools are used for risk assessment, but the most used ones are Ishikawa or fish-bone diagram for risk identification and failure mode and effect analysis for risk evaluation [36]. Risk assessment should be done in the first step of the drug product development, and analysis of these risk factors is recommended to be reconsidered during different stages of the drug development [37].

To establish the CMAs, the main components of liposomes are evaluated, that is the active substance, the lipids, and others like buffer solution [38]. Each active substance has different physiochemical properties and can influence the desired CQAs [9, 39]. Depending on its solubility, it can be entrapped in the aqueous core or in the lipid bilayer [40]. In order to achieve a greater therapeutic effect, more than one active substance may be encapsulated in the same product, so different physiochemical properties of one may influence the other/others drug entrapment process [40, 41]. Besides this, their octanol/buffer partition must also be taken into consideration because a low partition will lead to a prolonged release of the drug and vice versa [42].

Regarding the lipids, their physicochemical characteristics are also important. For example, lipids that contain unsaturated fatty acids are predisposed to degradation reactions like oxidation or hydrolysis and those which contain saturated fatty acids have a higher transition temperature (T_m) [43]. Another specific characteristic of lipids is their chain length. Usually, a shorter chain length will contribute to a thinner lipid bilayer and a larger internal volume for drug encapsulation. However, comparing three lipids with a difference of two carbon atoms between them, the lipid bilayer thickness obtained did not differ with more than 1 nm and a very small difference in liposomes' size and EE was observed [30]. Lipid properties can have a great impact on liposomes' membrane fluidity, permeability, or charge [44]. In this regard, cholesterol increases liposomal stability, reduces membrane fluidity, and, consequently, contributes to an increased EE [45].

Different compounds used for surface modification, in order to obtain a prolonged blood circulation or modified drug release, must be evaluated during risk identification and analysis. For example, when chitosan was used for coating, its concentration, solubility, and molecular weight influenced liposomal size [46]. In the case of polyethylene glycols (PEGs), the molecular weight and the density on liposomal surface influenced the biodistribution and size [47].

Besides the nature of liposomal components, their concentration and ratios between them are also critical. Some of these ratios, identified in risk assessments of different studies, are: organic phase-to-aqueous phase ratio; cholesterol-to-lecithin ratio; chloroform-to-methanol ratio; phospholipids-to-cholesterol molar ratio; and drug-to-lipid ratio [40, 46, 48, 49].

Regarding the preparation process, many techniques and methods like film hydration, emulsification, and reverse phase evaporation were developed but the most used one remains the film hydration method [37]. For this method to be efficient from the viewpoint of EE, parameters like temperature or rotation speed in the evaporation and hydration steps are critical and must be optimized [41, 49]. Film hydration method's great disadvantage is that the obtained liposomes are in the vast majority of micron size with a multilamellar structure, being characterized by a high percentage of lipids and a reduced internal volume. Thus, a reduction in particle size is mandatory, to increase their internal volume and to have a controlled size and narrow particle size distribution [30, 37]. For reduction of liposomal size, several techniques like sonication, freeze-thaw cycling, or extrusion may be used. Comparing these size-reduction methods through a screening experimental design, they were found to be critical for particle size distribution but also for EE [30]. When extrusion through membranes is used for size reduction, membrane pore size, temperature, and applied pressure are important CPPs because these factors define the final particle size and can seriously influence the final EE [30]. Sonication process also needs optimization as regards its time in order to achieve the desired liposomal size [46]. For the freeze-thaw cycling process, the number of the cycles might be taken into consideration for further optimization in order to get the desired particle size [30].

After the risk assessment, next step in QbD development is linking the identified CMAs and CPPs to CQAs by using Design of Experiments (DoE). The greatest advantage of using DoE is that it can track the interactions between the studied factors, CMAs and CPPs, and it can establish a quantitative relationship between the identified variables and the results [36]. For a better understanding of this concept in liposomes development, the most important CQAs will be discussed further, and how CMAs and CPPs can influence them according to different studies.

3.1 Drug content

The drug content is most frequently optimized in terms of EE than drug concentration as EE reflects better the preparation process performance and robustness. According to published data, the most influential factors for EE are: the amount of drug and phospholipids; cholesterol concentration; the nature of lipids and drug; the interactions between different components; the lipid-to-drug ratio; and several process parameters. Among these, several studies established, through DoE approaches, that lipid molar ratio and lipid-to-drug ratio are the most critical parameters for EE optimization. Using a great amount of lipids for liposome preparation favors the formation of many vesicles with a significant internal volume for drug encapsulation and, consequently, the EE of hydrophilic drugs increases [30, 45]. Including cholesterol in liposome formulations increases not only their stability but also the drug content, due to the so-called "pocket" theory, presuming that cholesterol can generate different size pockets inside the lipid bilayer where API can be entrapped [45]. The use of unsaturated lipids was shown to have a similar effect, the unsaturated lipids forming pockets inside the lipid bilayer where lipophilic drugs can be entrapped [50]. By increasing lipid-to-drug ratio from 1:1 to 10:1, the EE of ritonavir, a lipophilic drug, was doubled from 45 to 90% [43]. The ratio between phospholipid concentration and drug concentration was demonstrated to have a significant impact on EE in the case of simvastatin. Thus, at high simvastatin concentration and low concentration of phospholipids, EE decreased [51].

In the case of drugs with pH-dependent solubility like doxorubicin, the EE can be improved by the use of active loading method at specific pH range, the pH of hydrating buffer and of external buffer being critical parameters [52].

Regarding the drug concentration, several studies concluded that the total amount of drug used for encapsulation has a breaking point from which the EE cannot be further increased [41, 46].

EE may also be influenced by PPs. A fractional factorial design was used to establish the link between PPs like hydration temperature and the number of freeze-thaw cycles during preparation of liposomes loaded with FK50, an immunosuppressant. An interaction between these PPs was highlighted. It was observed that by using an increased number of cycles, and high hydration temperature, EE decreased, because repetition of the freeze-thaw cycle at high temperature might induce leakage of FK506 from the membrane of the liposomes [39]. In another study, it was observed that PPs like temperature during film hydration and rotation speed during solvent evaporation are critical. Using high temperature, above T_m, the concentration of the encapsulated drug increases because at high temperature, the lipid bilayer is more fluid and permits the entrapment of more drug. Also, using a high rotation speed at the film formation can have a great impact on EE because this CPP leads to formation of a thinner lipid bilayer which can easily be hydrated [53].

3.2 Particle size and size distribution

Maintaining a controlled particle size and low PDI is one of the main challenges when preparing liposomes, and optimization of these parameters is improved through the use of DoE. There are many critical factors influencing these parameters, both formulation and process related. For example, the influence of phospholipid concentration, active substances (curcumin and doxorubicin) concentration, working temperature, buffer pH, and phospholipid-to-cholesterol ratio, on liposomal size, was examined through a screening experimental design. Out of the studied factors, only phospholipid concentration and phospholipid-to-cholesterol ratio significantly influenced the size, while the concentration of the drugs and the working conditions were not critical for particle size. Noteworthy, none of the studied parameters influenced particle size distribution [41].

Regarding the effect of lipid concentration, different studies showed contradictory results on the size in relation to this factor [43, 46]. Usually, on increasing lipid concentration, liposomal size increases, simultaneously with PDI values [54]. Another observation was that increasing lipid concentration over a certain value leads to smaller size, probably due to lipid bilayer rearrangement into a bigger number of liposomes with smaller size and better size distribution [55]. Depending on the lipid type, a different influence on liposomal size was observed [56].

The active substances influence liposomal size depending on their physicochemical properties. It was observed that quercetin, a lipophilic compound, might replace some lipids in the lipid bilayer causing a reduction in size when it is used in high concentrations [55]. On the other hand, high concentrations of pravastatin, a hydrosoluble compound, led to a small size of liposomes due to an interaction between the active substance and the lipids [53].

3.3 Zeta potential

This parameter is influenced by formulation factors like ionic strength, pH, bilayer composition, or charged lipids and PPs like sonication time [48, 57]. In order to modulate the ZP values, different stabilizers, such as stearylamine and diacetyl phosphate, or modified lipids, such as poly(2-ethyl-2-oxazoline) (PEtOz) or PEG, can be incorporated in lipid bilayer. The concentration of these excipients may be optimized such as to obtain optimal stability [50, 57]. In conventional liposomes, cholesterol-to-lecithin ratio influences the zeta potential value [48]. Through DoE, several papers established the critical parameters influencing the zeta potential values. For example, a screening design study was used to determine which of the formulation factors (lipid concentration, cholesterol concentration, chitosan concentration, drug concentration, organic phase/aqueous phase ratio) and process parameters (temperature, stirring speed, sonication time) had a significant influence on zeta potential of chitosan-coated liposomes. As expected, chitosan concentration was a critical parameter, along with the temperature, which favored the coating process through reduction of vesicles size [46].

3.4 In vivo stability and drug release

By choosing a suitable lipid bilayer composition, a higher stability in vivo might be achieved. In order to sustain this, it was observed that using saturated lipids or cholesterol in formulations, lipid bilayer stability is increased and liposomal uptake by mononuclear phagocyte system will be reduced. Another used pathway is incorporation of different excipients, such as ganglioside GM1, phosphatidylinositol or PEG-lipids, creating a steric barrier which prevents their clearance from the system [42]. The organs in which liposomes accumulate for being eliminated are lungs, liver, spleen, and kidneys. It was observed that those liposomes with negative surface charge present a higher uptake in tumor cells and a slower uptake in liver cells [58]. In a study, four liposome formulations were compared regarding their *in vivo* stability, by tracking their accumulation in spleen. Results showed that the molecular weight of PEG attached to their surface as well as particle size were CMAs influencing the accumulation in the spleen [47].

Depending on the lipophilic or hydrophilic character of the active substance, the kinetics of release is different because the diffusion through liposomal membrane is influenced by its physiochemical properties [41]. This behavior was shown when the release of two different lipophilic drugs and a hydrophilic drug from the same liposomal system was studied in vitro. The lipophilic agent displayed prolonged release and a smaller total drug release in comparison with the hydrophilic one due to their different characteristics [40].

For pH-sensitive liposomes, the objective in terms of in vitro drug release is to have a very good stability of the encapsulated drug under physiological conditions and triggered drug release at certain pH values. To achieve this, the use of excipients having a membrane-destabilizing effect and their concentration are critical parameters [52, 56].

4. Defining the product and process design space

The design space (DS) is a multidimensional combination and interaction of the input variables, such as material attributes, and process parameters that have been shown to assure quality [9]. Thus, the advantage of DS determination is that it establishes the operating region which ensures consistent product qualities between different batches. Working within the DS, the product will meet the specifications mentioned in the QTPP, while moving out of it is considered a change that would normally require a regulatory approval in the case of authorized products [9].

Determination of the DS is based on multivariate analysis, considering the main effects of factors as well as their interaction, which helps in determining an operational region based on a predefined confidence level. The DS includes the product design space and process design space. The product DS is established with product's CQAs as dimensions, while the process design space is exhibited as CQAs with respect to CPPs [59].

One approach to establish the DS for liposomal products is to take simultaneously into consideration both formulation factors and CPPs. This method was used for the determination of the DS for lyophilized liposomes with simvastatin. Thus, the cholesterol concentration, the PEG proportion, the cryoprotectant-to-phospholipids molar ratio, and the number of extrusions through polycarbonate membranes were selected as the most influential factors for lyophilized liposome CQAs. Their variation range was determined, in which the established quality requirements of the product are met: reduced particle size, maximized drug retention during lyophilization, reduced change in phospholipid transition temperature, low residual moisture content. The validity of the DS was confirmed by determining the CQAs of a formulation corresponding to the robust set point, that is, the formulation for which the prediction errors are lowest. Thus, defining the DS was found to be a useful strategy for the development of stable lyophilized liposomes having predictable quality [60]. The same approach was used to generate the DS for preparation of prednisolon-loaded long-circulating liposomes at laboratory scale. In this case, the selected formulation parameters were prednisolon concentration and the PEG proportion in lipidic membrane, and the process parameters were the extrusion temperature and the rotation speed at the hydration of the lipid film. The DS was developed such as to ensure high liposomal drug concentration, high EE, and controlled size [49].

When process parameters are not found as critical for the quality of the product, or their impact on quality is easily controlled by fixing a certain operating level, the DS is proposed as a function of formulation variables. This approach has been used in several studies, the most studied formulation variables being the active substance concentration and the composition of the lipid membrane. For example, the DS for liposomes with tenofovir was constructed with respect to phospholipid, concentration, cholesterol concentration, and drug concentration, with a focus to obtain high drug encapsulation efficiency, as this was considered the most difficult property to predict and control for liposomes containing hydrophilic drugs. Other quality attributes considered critical in the study, particle size and stability, were controlled by the pore size of the extrusion membrane and by storing the samples at low temperature, respectively [61]. In another paper, the DS for chitosan-coated liposomes was established as a function of drug concentration, chitosan concentration, and the organic phase-to-aqueous phase ratio during liposome preparation by ethanol injection technique. These variables were found to be the only significant factors affecting the CQAs of the product, although other formulation and process parameters were evaluated through a screening study. The composite desirability function based on constraints was used to determine the conditions that would result in an optimal formulation design, in terms of particle size, encapsulation efficiency, and coating efficiency. Target values were selected for the mentioned quality attributes, and, on the basis of these target values, the optimum values for each variable or processing parameter were obtained. In an additional step, the robustness of the DS was analyzed and the results showed that the selected CPPs may help minimize the variations in QAs that might arise due to the variability of the raw materials [46].

The development of liposomal systems entrapping more than one active substances is more complex, because the properties of each will influence the CQAs of the product and their stability in the processing conditions will impact the process parameters. In this regard, a group of authors established and evaluated the DS for long-circulating liposomes co-encapsulated with curcumin and doxorubicin. DS development was based on a previous screening study, which revealed the critical parameters, that is, phospholipid concentration, the phospholipid-to-cholesterol molar ratio, doxorubicin concentration, and curcumin concentration. The purpose was to obtain the variation range of these factors for which the size of the liposomes is minimized and the encapsulation efficiencies of both drugs are maximized. The DS was established as the region within which the prediction of the CQAs is made with a probability of failure of less than 1%. Moreover, a DS hypercube was set out as a restricted zone in the DS where factors' values can vary independent of each other, without influencing the quality of the product [41].

The incomplete understanding of the manufacturing process is a major barrier in liposomal products' industrial production and clinical application. The destabilization of their structure during long-time storage as aqueous dispersions revealed the need for complex fabrication processes, involving drying steps such as lyophilization and spray drying. The key to the successful design and preparation of optimal liposomal dry powder formulations is an understanding of the significance of the drying process parameters [62]. This aim was achieved by several authors by determining the DS for lyophilization or spray drying process.

The DS for the freeze-drying process of pravastatin-loaded long-circulating liposomes was established as a function of the freezing rate and shelf temperature during primary drying. The two process parameters were found to have a great impact on product's CQAs, along with the presence of an annealing step. The conditions to obtain freeze-dried liposomes with the desired characteristics were generated using the combination trehalose-mannitol as cryoprotectant and by including an annealing step. A series of limitations and target values were applied for the critical quality attributes of the lyophilized product. Thus, the DS was constructed such as to ensure high drug retention after lyophilization, particle size below a certain value, low zeta potential, low residual moisture content, and a short duration of the primary drying. Out of the DS, an optimal formulation was selected and testing this formulation confirmed that the process delivers the desired quality of the product, as long as it is operated within the DS [63].

A process design space for spray drying of liposomes was developed such as to get a product that met the criteria for all CQAs of an inhalable powder formulation. When developing an inhalable product, the mass median aerodynamic diameter (MMAD) is the most important CQA, as this characteristic influences the deposition in the conducting airway. Besides this, the size stabilization, relative moisture content, and process yield were identified as CQAs. The DS was plotted by imposing restrictions especially for MMAD, which should be within the range 4.5–5.5 μm, but also for liposome size ratio (before/after drying) and the yield of the process, while for moisture content, no restrictions were set, as acceptably low values were obtained for all the experiments. First, an optimal operating space (OOS) was identified with a high feed flow rate, a low outlet temperature, a medium aspirator rate and in the area of low feedstock concentration and high atomizing airflow. The MMAD was the QA restricting the entire ODS to the low feedstock concentration and high atomizing airflow, whereas the other CQAs met the imposed criteria in a larger space [64].

5. The control strategy

Due to the great success of liposomal systems, not only in pharmaceutical formulation but also in cosmetics and food industry, there is a huge demand on developing and standardizing analytical and bioanalytical methods for liposome complete characterization, as well as for their detection in blood and tissues. Official guidance regarding the manufacturing and controls recommend several methods for lipid components as well as for the drug products [65]. According to FDA guideline, "liposome structures and integrity are important physicochemical properties and they reflect the ability of the liposome drug formulation to contain the drug substance and to retain the drug substance within the appropriate liposome structure" [65].

A key issue in the control of liposomes is closely related with the quality control of the lipid components, including modified lipids, which could dramatically influence the properties and performances of liposome drug product [65]. The quality of the final product is influenced by the source of lipids and also by the type of the lipids: synthetic, semi-synthetic, or natural.

All types of separation methods, such as gas chromatography (GC) or liquid chromatography (LC), capillary electrophoresis (CE), gel electrophoresis, or electrochromatography have proved their value in the analysis of lipid components of liposomes [66–68]. In order to evaluate the chemical stability of the liposomes, it is mandatory to assess the chemical stability of the lipid components in the final drug formulation, taking into account that some lipids could be degraded by oxidation or hydrolysis. From the practical point of view, most of the separation methods offer important information on this matter. The broad versatility, high selectivity, efficiency, and low time of analysis are making them a good choice.

While the analysis of lipids using liquid chromatography does not need a long stage of pretreatment, the lipids analyzed using gas chromatography have to be derivatized in order to obtain more volatile compounds which are not turning into degradation products at their boiling point [69]. For fatty acids, the most used derivatization method consists in the esterification of the acids and their transformation in methyl esters. The esterification takes place under various conditions of temperature, mixing process, using different catalysts and for various periods of time. There are plenty of studies among scientific literature about derivatization methods of fatty acids for their GC analysis, researchers being still interested in improving the pretreatment of these compounds to obtain an efficient and rapid process [70].

Even though the literature abounds in examples of separation techniques applied for lipid detection and quantification, some drawbacks are obvious: poor solubility, poor absorbance properties, the need for derivatization, and laborious sample preparations. Nevertheless, the hyphenation between techniques could overcome the drawbacks and offers promising results.

Regarding the control of the final product, the following properties are generally determined to characterize a liposome drug formulation: morphology; surface charge (expressed as zeta potential); particle size (average diameter) and polydispersity index; encapsulation efficiency; the amount of drug relative to the amount of lipids; phase transition; residual solvents; in vitro and in vivo drug release [71]. Variability in these properties may lead to changes in the quality of the liposomal drug products, including leakage of the drug from the liposomes. The QAs monitored for liposomes are presented in **Table 2**, together with the currently employed methods of analysis.

Particle size is one of the crucial parameters for further in vivo application of liposomes. The required size is usually in the range 20–250 nm. When using microscopy techniques, one will obtain a high-resolution three-dimensional profile of the vesicle surface under study. For instance, AFM permits liposomes' visualization without alteration of their native form, given that the requisite surface immobilization does not adversely affect the sample and that the force of the probe itself does not have deleterious effects on the vesicles [75]. AFM is a rapid, powerful, and relatively non invasive technique and compared to TEM, does not requiers complicated sample preparation and removal of liposomes from their native environment. Additionally, TEM provides information on the size distribution and shape of vesicles. Unfortunately, liposomes can suffer from structure perturbations triggered by the high vacuum conditions and the staining process.

The investigated property	Methods	Ref.
Morphology	UV-vis spectroscopy Spectrofluorimetry RMN Angle X-ray scattering (SAXS) Freeze fracture technique with subsequent transmission electron microscopy	[71–73]
Net charge (zeta potential)	ELS	[74]
Particle size	Microscopy techniques (TEM, AFM, SEM) SEC Field-flow fractionation and static DLS	[74, 75]
Drug encapsulation efficiency	Spectroscopy LC	[75]
In vitro drug release	Spectroscopy LC	[75]
In vivo drug release	Radiolabeling, fluorescence labeling, MRI, CT, MS	[75]

AFM, atomic force microscopy; TEM, transmission electron microscopy; SEM, scanning electron microscopy; ELS, electrophoretic light scattering; LC, liquid chromatography; SEC, size-exclusion chromatography; DLS, dynamic light scattering; MRI, magnetic resonance imaging; CT, computed tomography; MS, mass spectrometry.

Table 2.
The quality attributes monitored for liposomal products.

To measure particle size and size distribution, three types of techniques could be used: all together, separation, and particle-by-particle counting. In the first case, multi-angle (static) and dynamic light scattering (MALS and DLS, respectively) techniques permit the calculation of the average particle size and charge from the signal generated by multiple particles within the sample. Even though these techniques are fast, they have low resolution and do not provide the particle concentration. Separation techniques, such as disk centrifuge and field-flow fractionation, have the advantage of improving size analysis resolution by using differences in the sample properties, typically sedimentation rates, to pre-separate the sample prior to light-based (absorbance or scattering) analysis. The separation techniques do not measure particle concentration or charge, and they often suffer from separation-based issues. More effective are particle-by-particle counting techniques, such as tunable resistive pulse sensors (TRPSs). The main advantage is the possibility to measure the properties of individual liposomes, offering a direct measurement of the particle concentration as well as high resolution and more accurate analysis of the particle size and charge (zeta potential) distribution. This ability to simultaneously measure the distribution of both the size and zeta potential represents a new and effective means of analyzing liposome properties [76].

To measure the zeta potential, ELS technique is currently used. It consist in using heterodyne scattering methods in which a fraction of the laser beam is split away by a mirror before reaching the sample, and is directed to the detector where it is combined with scattered light from particles diffusing in the sample. The fraction of redirected light is referred to as the "local oscillator" and, unlike the light scattered by the sample, does not fluctuate. It is used as a reference beam and must be much larger than the average intensity of the scattered light produced by the diffusing particles. To determine the zeta potential, the electrophoretic mobility must first be ascertained [77].

The surface modification of liposomes is sometimes performed in order to increase their in vivo drug-delivery performances. As mentioned earlier, several ways of modifying the surfaces were reported, like the addition of polyethylene glycol (PEG) chains, or the attachment of antibodies and cellular receptor recognition molecules (e.g., the RGD peptide) as molecular targeting probes. Tracking the successful modification of liposomes is made by measuring the change in their electrophoretic mobility, in fact the modification of their zeta potential arising from the change in the number of charged surface groups. Another method for measuring the zeta potential of liposome particles is via ensemble light scattering techniques, which use a similar principle as DLS [78].

Within the control strategy, process analytical technology (PAT) has been evidenced lately as a significant tool for measuring parameters and attributes related to the active substance, the finished product as well as the processing conditions. Among the PAT tools, in the field of liposomes, near-infrared spectroscopy (NIRs) has been shown to be useful for the chemical characterization of liposomes in terms of the composition of the lipid membrane, as well as for the simultaneous quantification of excipients and active substance [79, 80].

Besides the large-scale production, which is a major challenge, the standardization of analysis procedures easily scalable is another important goal in practical application of liposomes. The liposome production is done in small size batches compared with other pharmaceutical products which are produced in large batches [71]. However, the possible application of liposomes and the increasing number of clinical trials involving liposomes prove the fact that this field of research is very dynamic and the synthesis and analysis methods become more effective, encouraging their application in the development of new drug carriers.

6. Conclusions

The implementation of the QbD approach in the pharmaceutical industry is intended to enhance the quality of pharmaceutical products through identifying, analyzing, and controlling all factors that could alter their quality, and, consequently, its efficacy and safety. Currently, an increasing number of papers describe the development of liposomal products based on this risk-based approach, although only few are following all the steps recommended by the regulatory guidelines. Finally, this strategy could be useful to promote liposomes from laboratory into authorized products.

Acknowledgements

The authors are grateful for the financial support from the Romanian National Authority for Scientific Research and Innovation, CNCS/CCCDI-UEFISCDI; project number PN-III-P1-1.2-PCCDI2017-022159PCCDI/2018 (IMPROVE), within PNCDI III.

Conflict of interest

The authors declare that there is no conflict of interest.

Author details

Alina Porfire*, Marcela Achim, Cristina Barbalata, Iulia Rus, Ioan Tomuta and Cecilia Cristea
Faculty of Pharmacy, University of Medicine and Pharmacy "Iuliu Hatieganu", Cluj Napoca, Romania

*Address all correspondence to: aporfire@umfcluj.ro

References

[1] Rayaprolu BM. Quality by design: A brief introduction. Journal of Pharmacovigilance. 2015;**3**(4):e142

[2] Tefas LR, Rus LM, Achim M, Vlase L, Tomuta I. Application of the quality by design concept in the development of quercetin-loaded polymeric nanoparticles. Farmácia. 2018;**66**(5):798-810

[3] Sangshetti JN, Deshpande M, Zaheer Z, Shinde DB, Arote R. Quality by design approach: Regulatory need. Arabian Journal of Chemistry. 2017;**10**:S3412-S3425

[4] Lionberger RA, Lee SL, Lee L, Raw A, Yu LX. Quality by design: Concepts for ANDAs. The AAPS Journal. 2008;**10**(2):268-276

[5] Yu LX, Amidon G, Khan MA, Hoag SW, Polli J, Raju GK, et al. Understanding pharmaceutical quality by design. The AAPS Journal. 2014;**16**(4):771-783

[6] The International Council for Harmonisation of Technical Requirements for Pharmaceuticals for Human Use (ICH) Guidelines [Internet]. 2019. Available from: https://www.ich.org/products/guidelines/quality/article/quality-guidelines.html

[7] FDA Guidances [Internet]. 2019. Available from: https://www.fda.gov/Drugs/GuidanceComplianceRegulatoryInformation/Guidances%20/default.htm

[8] EMA Reflection Paper: Chemical, pharmaceutical and biological information to be included in dossiers when Process Analytical Technology (PAT) is employed [Internet]. 2006. Available from: http://www.ema.europa.eu/docs/en_GB/document%20_library/Other/2009/10/WC500004890.pdf

[9] ICH harmonised tripartite guideline. Pharmaceutical development Q8(R2). 2009. Available from: https://www.ich.org/fileadmin/Public_Web_Site/ICH_Products/Guidelines/Quality/Q8_R1/Step4/Q8_R2_Guideline.pdf [Accessed 01-2019]

[10] Raman NVVSS, Mallu UR, Bapatu HR. Analytical quality by design approach to test method development and validation in drug substance manufacturing. Journal of Chemistry. 2015;**2015**:1-8

[11] Pramod K, Tahir MA, Charoo NA, Ansari SH, Ali J. Pharmaceutical product development: A quality by design approach. International Journal of Pharmaceutical Investigation. 2016;**6**(3):129-138

[12] Chowdary KPR, Shankar KR, Kumar PS. Recent research on QbD approach in formulation development – A review. International Journal of Computer Science Trends and Technology. 2014;**4**(1):282-292

[13] Yu LX. Pharmaceutical quality by design: Product and process development, understanding, and control. Pharmaceutical Research. 2008;**25**(4):781-791

[14] Sylvester B, Porfire A, Muntean DM, Vlase L, Banciu M, Luput L, et al. Optimzation of prednisolone-loaded long-circulating liposomes via application of Quality by Design (QbD) approach. Journal of Liposome Research. 2018;**28**(1):49-61

[15] Shirsat AE, Chitlange SS. Application of quality by design approach to optimize process and formulation parameters of rizatriptan loaded chitosan nanoparticles. Journal of Advanced Pharmaceutical Technology & Research. 2015;**6**(3):88-96

[16] De Beer TR, Wiggenhorn M, Hawe A, Kasper JC, Almeida A, Quinten T,

et al. Optimization of a pharmaceutical freeze-dried product and its process using an experimental design approach and innovative process analyzers. Talanta. 2011;**83**(5):1623-1633

[17] Eon-Duval A, Valax P, Solacroup T, Broly H, Gleixner R, Strat CL, et al. Application of the Quality by Design approach to the drug substance manufacturing process of an Fc fusion protein: Towards a global multi-step design space. Journal of Pharmaceutical Sciences. 2012;**101**(10):3604-3618

[18] Sangshetti JN, Deshpande M, Zaheer Z, Shinde DB, Arote R. Quality by design approach: Regulatory need. Arabian Journal of Chemistry. 2014;**101**:1-14

[19] Mishra V, Thakur S, Patil A, Shukla A. Quality by design (QbD) approaches in current pharmaceutical set-up. Expert Opinion on Drug Delivery. 2018;**15**(8):737-758

[20] Xu X, Khan MA, Burgess DJ. A quality by design (QbD) case study on liposomes containing hydrophilic API: II. Screening of critical variables, and establishment of design space at laboratory scale. International Journal of Pharmaceutics. 2012;**423**:543-553

[21] Beg S, Rahman M, Kohli K. Quality-by-design approach as a systematic tool for the development of nanopharmaceutical products. Drug Discovery Today. 2019;**24**(3):717-725

[22] Rizvi SAA, Saleh AM. Applications of nanoparticle systems in drug delivery technology. Saudi Pharmaceutical Journal. 2018;**26**:64-70

[23] Gabizon AA. Stealth liposomes and tumor targeting: one step further in the quest for the magic bullet. Clinical Cancer Research. 2001;7:223-225

[24] European Pharmacopoeia Online 9.3 [Internet]. 2018. Available from: http://online6.edqm.eu/ep903/ [Accessed 01-02-2019]

[25] Sesarman A, Tefas L, Sylvester B, Licarete E, Rauca V, Luput L, et al. Co-delivery of curcumin and doxorubicin in PEGylated liposomes favored the antineoplastic C26 murine colon carcinoma microenvironment. Drug Delivery and Translational Research. 2019;**9**(1):260-272

[26] Gaumet M, Vargas A, Gurny R, Delie F. Nanoparticles for drug delivery: The need for precision in reporting particle size parameters. European Journal of Pharmaceutics and Biopharmaceutics. 2008;**69**:1-9

[27] Fukumori Y, Ichikawa H. Nanoparticles for cancer therapy and diagnosis. Advanced Powder Technology. 2006;**17**(1):1-28

[28] Aftab S, Shah A, Nadhman A, Kurbanoglu S, Ozkan SA, Dionysiou DD, et al. Nanomedicine: An effective tool in cancer therapy. International Journal of Pharmaceutics. 2018;**540**:132-149

[29] Bertrand N, Wu J, Xu X, Kamaly N, Farokhzad OC. Cancer nanotechnology: The impact of passive and active targeting in the era of modern cancer biology. Advanced Drug Delivery Reviews. 2014;**66**:2-25

[30] Xu X, Khan MA, Burgess DJ. A quality by design (QbD) case study on liposomes containing hydrophilic API: I. Formulation, processing design and risk assessment. International Journal of Pharmaceutics. 2011;**419**:52-59

[31] Amasya G, Badilli U, Aksu B, Tarimci N. Quality by design case study 1: Design of 5-fluorouracil loaded lipid nanoparticles by the W/O/W double emulsion—Solvent evaporation method. European Journal of Pharmaceutical Sciences. 2016;**84**:92-102

[32] He K, Tang M. Safety of novel liposomal drugs for cancer treatment:

Advances and prospects. Chemico-Biological Interactions. 2018;**295**:13-19

[33] Li M, Du C, Guo N, Teng Y, Meng X, Sun H, et al. Composition design and medical application of liposomes. European Journal of Medicinal Chemistry. 2019;**164**:640-653

[34] Allen TM, Cullis PR. Liposomal drug delivery systems: From concept to clinical applications. Advanced Drug Delivery Reviews. 2013;**65**:36-48

[35] ICH harmonised tripartite guideline. Quality Risk Management Q9 [Internet]. 2005;(November):1-23. Available from: http://scholar.google.com/scholar?hl=en&btnG=Search&q=intitle:Quality+Risk+Management+Q9#0%5Cnhttp://www.cls.co.at/downloads/ich-q9---quality-risk-management.pdf [Accessed 01-2019]

[36] Zhang L, Mao S. Application of quality by design in the current drug development. Asian Journal of Pharmaceutical Sciences. 2017;**12**(1):1-8

[37] Simões A, Veiga F, Figueiras A, Vitorino C. A practical framework for implementing Quality by Design to the development of topical drug products: Nanosystem-based dosage forms. International Journal of Pharmaceutics. 2018;**548**(1):385-399

[38] U.S. Department of Health and Human Services Food and Drug Administration Center for Drug Evaluation and Research. Guidance for industry: Liposome drug products: chemistry, manufacturing, and controls; human pharmacokinetics and bioavailability; and labeling documentation. Pharmaceutical Quality [Internet]. 2018. Available from: https://www.fda.gov/Drugs/GuidanceComplianceRegulatoryInformation/Guidances/ucm064979.htm

[39] Toyota H, Asai T, Oku N. Process optimization by use of design of experiments: Application for liposomalization of FK506. European Journal of Pharmaceutical Sciences. 2017;**102**:196-202

[40] Lin J, Wang X, Wu Q, Dai J, Guan H, Cao W, et al. Development of Salvianolic acid B-Tanshinone II A-Glycyrrhetinic acid compound liposomes: Formulation optimization and its effects on proliferation of hepatic stellate cells. International Journal of Pharmaceutics. 2014;**462**(1-2):11-18

[41] Tefas LR, Sylvester B, Tomuta I, Sesarman A, Licarete E, Banciu M, et al. Development of antiproliferative long-circulating liposomes co-encapsulating doxorubicin and curcumin, through the use of a quality-by-design approach. Drug Design, Development and Therapy. 2017;**11**:1605-1621

[42] Abu Lila AS, Ishida T. Liposomal delivery systems: Design optimization and current applications. Biological & Pharmaceutical Bulletin. 2017;**40**(1):1-10

[43] Patel GM, Shelat PK, Lalwani AN. QbD based development of proliposome of lopinavir for improved oral bioavailability. European Journal of Pharmaceutical Sciences. 2017;**108**:50-61. DOI: 10.1016/j.ejps.2016.08.057

[44] Zununi Vahed S, Salehi R, Davaran S, Sharifi S. Liposome-based drug co-delivery systems in cancer cells. Materials Science and Engineering: C. 2017;**71**:1327-1341

[45] Xu X, Costa AP, Khan MA, Burgess DJ. Application of quality by design to formulation and processing of protein liposomes. International Journal of Pharmaceutics. 2012;**434**(1-2):349-359

[46] Pandey AP, Karande KP, Sonawane RO, Deshmukh PK. Applying quality by design (QbD) concept for fabrication of

chitosan coated nanoliposomes. Journal of Liposome Research. 2014;**24**(1):37-52

[47] Sieber S, Grossen P, Uhl P, Detampel P, Mier W, Witzigmann D, et al. Zebrafish as a predictive screening model to assess macrophage clearance of liposomes in vivo. Nanomedicine. 2019;**17**:82-93

[48] Sahu AK, Jain V. Screening of process variables using Plackett–Burman design in the fabrication of gedunin-loaded liposomes. Artificial Cells, Nanomedicine, and Biotechnology. 2017;**45**(5):1011-1022

[49] Sylvester B, Porfire A, Muntean DM, Vlase L, Lupuţ L, Licarete E, et al. Optimization of prednisolone-loaded long-circulating liposomes via application of Quality by Design (QbD) approach. Journal of Liposome Research. 2018;**28**(1):49-61

[50] Ingle SG, Pai RV, Monpara JD, Vavia PR. Liposils: An effective strategy for stabilizing Paclitaxel loaded liposomes by surface coating with silica. European Journal of Pharmaceutical Sciences. 2018;**122**:51-63

[51] Porfire A, Tomuta I, Muntean D, Luca L, Licarete E, Alupei MC, et al. Optimizing long-circulating liposomes for delivery of simvastatin to C26 colon carcinoma cells. Journal of Liposome Research. 2015;**25**(4):261-269

[52] Rehman AU, Omran Z, Anton H, Mély Y, Akram S, Vandamme TF, et al. Development of doxorubicin hydrochloride loaded pH-sensitive liposomes: Investigation on the impact of chemical nature of lipids and liposome composition on pH-sensitivity. European Journal of Pharmaceutics and Biopharmaceutics. 2018;**133**(74):331-338

[53] Sylvester B, Porfire A, Muntean DM, Vlase L, Tomuţă I. Formulation optimization of pravastatin loaded long-circulating liposomes using a design of experiments. Farmácia. 2016;**64**(3):449-458

[54] Achim M, Tomuţă I, Muntean D, Porfire A, Tefas LR, Patras L, et al. Optimization and in vitro evaluation of 5-fluorouracil—loaded long—circulating liposomes. Farmácia. 2017;**65**(1):82-91

[55] Tefas LR, Muntean DM, Vlase L, Porfire AS, Achim M, Tomuţă I. Quercetin-loaded liposomes: Formulation optimization through a D-optimal experimental design. Farmácia. 2015;**63**(1):26-33

[56] Xu H, Hu M, Yu X, Li Y, Fu Y, Zhou X, et al. Design and evaluation of pH-sensitive liposomes constructed by poly(2-ethyl-2-oxazoline)-cholesterol hemisuccinate for doxorubicin delivery. European Journal of Pharmaceutics and Biopharmaceutics. 2015;**91**(3):66-74

[57] Smith MC, Crist RM, Clogston JD, McNeil SE. Zeta potential: a case study of cationic, anionic, and neutral liposomes. Analytical and Bioanalytical Chemistry. 2017;**409**(24):5779-5787

[58] Agrahari V, Agrahari V. Facilitating the translation of nanomedicines to a clinical product: challenges and opportunities. Drug Discovery Today. 2018;**23**(5):974-991

[59] Li J, Qiao Y, Wu Z. Nanosystem trends in drug delivery using quality-by-design concept. Journal of Controlled Release. 2017;**256**:9-18

[60] Porfire A, Muntean DM, Rus L, Sylvester B, Tomuta I. A quality by design approach for the development of lyophilized liposomes with simvastatin. Saudi Pharmaceutical Journal. 2017;**25**:981-992

[61] Xu X, Khan MA, Burgess DJ. A quality by design (QbD) case study on liposomes hydrophilic API: II. Screening

of critical variables, and establishment of design space at laboratory scale. International Journal of Pharmaceutics. 2012;**423**:543-553

[62] Ingvarsson PT, Yang M, Nielsen HM, Rantanen J, Foged C. Stabilization of liposomes during drying. Expert Opinion on Drug Delivery. 2011;**8**(3):375-388

[63] Sylvester B, Porfire A, Achim M, Rus L, Tomuta I. A setp forward towards the development of stable freeze-dried liposomes: a quality by design approach (QbD). Drug Development and Industrial Pharmacy. 2018;**44**(3):385-397

[64] Ingvarsson PT, Yang M, Mulvad H, Nielsen HM, Rantanen J, Foged C. Engineering of an inhalable DDA/TDB liposomal adjuvant: a Quality by Design approach towards optimization of the spray drying process. Pharmaceutical Research. 2013;**30**:2772-2784

[65] Center for Drug Evaluation and Research, Food and Drug Administration. Liposome Drug Products. Guidance for Industry [Internet]. 2018. Available from: https://www.fda.gov/downloads/drugs/guidances/ucm070570.pdf

[66] Hübschmann HJ. Handbook of GC/MS Fundamentals and Applications. 2nd ed. Weinheim: Wiley-VCH; 2008. https://www.wiley.com/en-us/Handbook+of+GC+MS%3A+Fundamentals+and+Applications%2C+3rd+Edition-p-9783527334742

[67] Fanali S, Haddad PR, Poole C, Riekkola M-L. Liquid Chromatography. Volume 2, Applications. 2nd ed. Amsterdam: Elsevier; 2017

[68] Landers JP. Handbook of capillary and microchip electrophoresis and associated microtechniques. 3rd ed. Boca Raton: CRC Press; 2007. https://www.crcpress.com/Handbook-of-Capillary-and-Microchip-Electrophoresis-and-Associated-Microtechniques/Landers/p/book/9780849333293

[69] Dettmer-Wilde K, Engewald W. Practical Gas Cromatography. Heidelberg: Springer; 2014

[70] Saliu F, Orlandi M. In situ alcoholysis of triacylglycerols by application of switchable-polarity solvents. A new derivatization procedure for the gas-chromatographic analysis of vegetable oils. Analytical and Bioanalytical Chemistry. 2013;**405**:8677-8684

[71] Laouini A, Sfar S, Charcosset C, Fessi H, Lyon CPE, Bat G, et al. Preparation, characterization and applications of liposomes: State of the art. Journal of Colloid Science and Biotechnology. 2012;**1**(2):147-168

[72] Subramanian A, Velander WH. Dot-blot analysis of the degree of covalent modification of proteins and antibodies at amino groups. Journal of Immunological Methods. 1997;**203**:45-53

[73] Mukherjee S, Raghuraman H, Dasgupta S, Chattopadhyay A. Organization and dynamics of N - (7-nitrobenz-2-oxa- 1 , 3-diazol-4-yl) -labeled lipids: a fluorescence approach. Chemistry and Physics of Lipids. 2004;**127**:91-101

[74] Baxa U. Imaging of liposomes by transmission electron microscopy. In: Characterization of Nanoparticles Intended for Drug Delivery Methods in Molecular Biology. New York: Human Press; 2018

[75] Teschke O. Liposome structure imaging by atomic force microscopy: Verification of improved liposome stability during adsorption of multiple

aggregated vesicles. Langmuir. 2002;**18**(17):6513-6520

[76] Kozak D, Anderson W, Vogel R, Chen S, Antaw F, Trau M. Simultaneous size and zeta-potential measurements of individual nanoparticles in dispersion using size-tunable pore sensors. ACS Nano. 2012;**6**(8):6990-6997

[77] Electrophoretic Light Scattering (ELS) [Internet]. 2019. Available from: https://www.malvernpanalytical.com/en/products/technology/light-scattering/electrophoretic-light-scattering?utm_source=MaterialsTalks&utm_medium=Blog&utm_campaign=Liposome&utm_term=22724

[78] Ying, B. and R. Campbell, Polyethylene-glycol (PEG) enhances siRNA delivery and reduces toxic effect of cationic lipids under optimized experimental conditions. Proceedings of the American Association for Cancer Research Annual Meeting; 2009;**50**:153-154

[79] Porfire A, Tomuta I, Tefas L, Leucuta SE, Achim M. Simultaneous quantification of L-alpha-phosphatidylcoline and cholesterol in liposomes using near infrared spectrometry and chemometry. Journal of Pharmaceutical and Biomedical Analysis. 2012;**63**:87-94

[80] Porfire A, Muntean D, Achim M, Vlase L, Tomuta I. Simultaneous quantification of simvastatin and excipients in liposomes using near infrared spectroscopy and chemometry. Journal of Pharmaceutical and Biomedical Analysis. 2015;**107**:40-49